# SOIL DEGRADATION
## in the
# UNITED STATES

*Extent, Severity, and Trends*

# SOIL DEGRADATION
## in the
# UNITED STATES
### *Extent, Severity, and Trends*

Rattan Lal
Terry M. Sobecki
Thomas Iivari
John M. Kimble

CRC Press
Taylor & Francis Group
Boca Raton London New York

CRC Press is an imprint of the
Taylor & Francis Group, an **informa** business

First published 2004 by Lewis Publishers

Published 2020 by CRC Press
Taylor & Francis Group
6000 Broken Sound Parkway NW, Suite 300
Boca Raton, FL 33487-2742

© 2004 by Taylor & Francis Group, LLC
CRC Press is an imprint of Taylor & Francis Group, an Informa business

No claim to original U.S. Government works

ISBN 13: 978-0-367-57846-6 (pbk)
ISBN 13: 978-1-56670-534-9 (hbk)

**Visit the Taylor & Francis Web site at**
**http://www.taylorandfrancis.com**

**and the CRC Press Web site at**
**http://www.crcpress.com**

## Library of Congress Cataloging-in-Publication Data

Lal, Rattan
   Soil degradation in the United States: extent, severity, and trends/ Rattan Lal, Terry M. Sobecki, Thomas Iivari, John M. Kimble.
      p. cm.
   Includes bibliographical references and index.
   ISBN 1-56670-534-7
      1. Soil degradation–United States, I. Sobecki, Terry M. II. Iivari, Thomas III. Kimble, John M. IV. Title.

S624.A1.S63 2003
333.76'137'0973–dc21                                            2003044233

# Foreword

Hugh Hammond Bennet, at an address given in 1940, lamented, "I have never been able to understand why, with rivers running red or yellow with every heavy rain ..., and with new tributaries in the form of gullies being cut across thousands of fields, so few people thought to relate these obvious facts of land attrition to a rapidly increasing menace to American agriculture." He concluded his talk with the statement, "Back of all human security must be the security of our physical resources, because both human opportunity and security, in the final analysis, are built on the permanent productivity of the land."

Water erosion specifically and soil degradation in general are continuing threats with negative impacts not only on agriculture but also on the environment. Water and air quality are directly or indirectly affected by activities on land and the assessment and monitoring of all forms of degradation are essential to evaluate trends and implement mitigating technologies.

This book on soil degradation in the United States is an assessment of the current trends and geographic hot spots. It not only describes the kinds of soil degradation but also evaluates the processes and the triggers responsible. By evaluating each kind of degradation in its geographical and land management context, the authors have presented information useful to a wide range of land users.

The soil resource, like other resources, is being stressed and society must become more aware of this and play a more integral role in their protection and conservation. Initially an economic and environmental issue, it is becoming a moral issue that is calling for action.

**Maurice J. Mausbach**
*Deputy Chief, Soil Survey and Resource Assessment*
*USDA Natural Resources Conservation Service*

# Preface

Soil degradation is one of the important environmental issues of the 21st century. Its importance is accentuated by its actual and potential impacts on biomass productivity, water and air quality, and emission of greenhouse gases into the atmosphere. Soil degradation affects biomass productivity by creating water (drought or anaerobiosis) and elemental imbalance (deficiency or toxicity) in the root zone, decreasing effective root depth, and increasing susceptibility to pests. Soil degradation affects water quality by transport of suspended and dissolved loads in surface water, and of agricultural chemicals into ground water. Nonpoint source pollution, accentuated by soil degradation, is a principal cause of eutrophication of surface and contamination of ground water. Soil degradation impacts climate change directly and indirectly. Directly, it leads to emission of greenhouse gases (GHGs) by increasing mineralization of soil organic matter under aerobic ($CO_2$) and anaerobic ($CH_4$) conditions. Soil degradation also increases emission of $N_2O$ from denitrification. Indirectly, soil degradation decreases biomass productivity with the attendant adverse impacts on the quality and the quantity of the biomass returned to the soil.

There is a strong need for easy access to credible data on the extent and severity of soil degradation by different processes. It is difficult to assess reliability and the precise environmental and economic impacts of soil degradation without knowing the dimensions (extent, magnitude, and rate) of soil degradation by different processes. Decisions made on land use conversion and other management options made on sketchy or unreliable data can lead to costly and gross errors. Erroneous and unreliable information is worse than having no information because it can lead to erroneous decisions on critical policy issues.

This book is an attempt to make information readily available on soil degradation by different processes in the U.S. The book is comprised of 14 chapters divided into four sections. Section I, the first four chapters, deals with basic processes. Definitions, types of soil degradation, factors and causes of soil degradation, soil quality, and soil resilience are discussed in Chapter 1. Chapter 2 describes the global extent of soil degradation including the Global Assessment of Soil Degradation (GLASOD) methodology. Chapter 3 describes predominant land uses in the U.S., especially the historical trends in cropland and grazing land. Chapter 4 outlines the soil degradation category indicators methodology used in the book. Section II is comprised of five chapters dealing with soil degradation by water and wind erosion. Soil erosion by water from cropland is described in Chapter 5, and erosion by water from other lands (grazing land, CRP land, and minor lands) is examined in Chapter 6. Wind erosion is discussed in Chapter 7, water and wind erosion on cropland in Chapter 8, and total erosion on pasture and CRP lands in Chapter 9. The four chapters comprising Section III deal with other forms of soil degradation in the U.S. The extent and severity of desertification is described in Chapter 10, salinization and other salt-affected soils in Chapter 11, soil degradation by drastic disturbance including mining in Chapter 12, and degradation of wetlands in Chapter 13. Section IV has one chapter describing policy options related to soil degradation. The book also provides four appendices including a table of conversion of units.

The data in this book are a result of analyses of previously published information. However, we feel it is necessary and proper to give credit to the sources, since many analyses of our natural resources conditions and trends would be prohibitively expensive and unattainable for all except the largest organizations and institutions. Much of the information developed in this report was a result of analysis of the Natural Resources Conservation Service's (NRCS) National Resources Inventory (NRI) data.

The NRI is a scientifically based, statistically valid inventory of the status, condition, and trend of aspects of the soil and water resources on private land throughout the U.S., which is focused on conserving our nation's soil resource, maintaining agricultural productivity, and preventing degradation of water quality. The sample design used in the NRI enables area-based estimates of soil and land conditions and trends that are statistically valid at the Major Land Resource Area level (i.e., natural geographic areas that cover large portions of multi-state regions of the U.S.).

At the time of this work's preparation, two of the authors (Terry Sobecki and Thomas Iivari) were on the staff of the USDA-NRCS. We would like to thank our colleagues in the Resource Inventory Division and the Resources Assessment Division of the NRCS for their help and assistance during the preparation of this work.

Ancillary data were also used in this report, including estimates of sediment delivery developed by an expert panel of NRCS sedimentation geologists by Agricultural Producing Areas in the U.S. for use in the Resource Conservation Act (RCA) appraisal process. Agricultural Producing Areas are essentially county-normalized subbasins of major river basins in the U.S. The Agricultural Producing Areas were originally established by the Iowa State University Center for Agricultural and Rural Development (CARD) for use in agricultural resource and policy analysis. Some of these data will also be presented in a companion volume on soil restoration.

We have made every effort to use site-specific information from the above data sources in the text. Again, we wish to acknowledge the important role that availability of such data resources plays in taking new or unique looks at the state of our nation's natural resource base.

Thanks are due to the staff of CRC Press, especially Erika Dery, for her efforts in facilitating the publication of this book and making it available to the scientific and policy communities. We are also thankful to Jeremy Alder and Pat Patterson of The Ohio State University for their help in preparing several graphs. Photographs depicting different types of soil degradation processes have been obtained from numerous sources, including the USDA. Some photographs (depicting erosion, grazing, and urban encroachment) were made available by the staff (especially Jodi Miller) of the Communication and Technology Department of the College of Food, Agriculture and Environmental Science of The Ohio State University. We offer special thanks to Ms. Brenda Swank of The Ohio State University for typing the text and organizing the material to get it ready for publication. The efforts of many others were very important in publishing this relevant and important work.

**R. Lal**
**T. Sobecki**
**T. Iivari**
**J.M. Kimble**

# The Authors

**Rattan Lal**, Ph.D., is a professor of soil science in the School of Natural Resources and director of the Carbon Management and Sequestration Center, at The Ohio State University. Prior to joining Ohio State in 1987, he served as a soil scientist for 18 years at the International Institute of Tropical Agriculture, Ibadan, Nigeria. In Africa, Professor Lal conducted long-term experiments on soil erosion processes as influenced by rainfall characteristics, soil properties, methods of deforestation, soil-tillage and crop-residue management, cropping systems including cover crops and agroforestry, and mixed or relay cropping methods. He also assessed the impact of soil erosion on crop yield and related erosion-induced changes in soil properties to crop growth and yield.

Since joining OSU, he has continued research on erosion-induced changes in soil quality and developed a new project on soils and climate change. He has demonstrated that accelerated soil erosion is a major factor affecting emission of carbon from soil to the atmosphere. Soil-erosion control and adoption of conservation–effective measures can lead to carbon sequestration and mitigation of the greenhouse effect. Other research interests include soil compaction, conservation tillage, minesoil reclamation, water table management, and sustainable use of soil and water resources of the tropics.

Professor Lal is a fellow of the Soil Science Society of America, American Society of Agronomy, Third World Academy of Sciences, American Association for the Advancement of Science, Soil and Water Conservation Society, and Indian Academy of Agricultural Sciences. He is the recipient of the International Soil Science Award, the Soil Science Applied Research Award, and Soil Science Research Award of the Soil Science Society of America, the International Agronomy Award of the American Society of Agronomy, and the Hugh Hammond Bennet Award of the Soil and Water Conservation Society. Dr. Lal has been awarded an honorary degree of Doctor of Science from Punjab Agricultural University, India. He is a past president of the World Association of Soil and Water Conservation and the International Soil Tillage Research Organization. He was a member of the U.S. National Committee on Soil Science of the National Academy of Sciences (1998–2002), and he has served on the Panel on Sustainable Agriculture and the Environment in the Humid Tropics of the National Academy of Sciences.

**Terrence M. Sobecki** is a soil scientist with more than 20 years experience in natural resources and environmental applications of soil science. Dr. Sobecki started his career as a soil scientist with the Natural Resources Conservation Service (formerly Soil Conservation Service), working as a field soil surveyor for the National Cooperative Soil Survey (NCSS) program. While on the staff of the NRCS Soil Survey Laboratory he was responsible for conducting soil survey investigations supporting NCSS soil mapping and interpretations programs. As a water quality specialist with NRCS he provided technical assistance to NRCS offices and programs in the western U.S. in the areas of nutrient and contaminant fate and transport and received numerous awards and recognition. He spent 4 years with the NRCS

Resources Assessment Division in the agency national headquarters performing resource assessments and analyses in the areas of rangeland health and soil quality.

Dr. Sobecki is currently chief of the Environmental Processes Branch of the U.S. Army Corps of Engineers Engineer Research and Development Center's Environmental Laboratory. He supervises a multidiscipline staff conducting research in the areas of water quality, contaminant fate and transport in terrestrial and aquatic systems, and invasive plant species control. He also conducts research in soil processes impacting contaminated soil cleanup and remediation.

**Thomas Iivari** is a dedicated professional with a wide range of experience with natural resources issues. He started his career as a geologist with the Ohio Department of Natural Resources, and shortly thereafter became New Jersey state geologist for the Natural Resources Conservation service (formerly Soil Conservation Service). For 15 years Mr. Iivari served as northeast regional geologist and water resources planning specialist with the NRCS Northeast National Technical Center, providing geological and planning technical assistance to NRCS field and state offices in the northeastern region of the U.S. Mr. Iivari also served as NRCS national geologist and as senior hydrologist at the agency headquarters in Washington, D.C. where he served as senior technical specialist and policy advisor on a range of natural resource issues, including clean water and animal waste programs. Mr. Iivari received numerous awards and recognition for his contributions to natural resource conservation while with NRCS. He also served as ASTM vice chairman, Committee on Water–Subcommittee on Sedimentation and Hydrology. Mr. Iivari now works with the U.S. EPA in the northeastern U.S. where he continues to apply his expertise in geology and water resources to protecting our nation's natural resources.

**John M. Kimble**, Ph.D., is a research soil scientist at the USDA Natural Resources Conservation Service, National Soil Survey Center in Lincoln, Nebraska, where he has been for the last 21 years. Previously he served as a field soil scientist in Wyoming for 3 years and as an area soil scientist in California for 3 years. He has received the International Soil Science Award from the Soil Science Society of America. While in Lincoln, he worked on a U.S. Agency for International Development Project for 11 years, helping developing countries with their soil resources, and he remains active in international activities. For the last 10 years he has focused more on global climate change and the role soils can play in this area. His scientific publications deal with topics related to soil classification, soil management, global climate change, and sustainable development. He has worked in many different ecoregions, from the Antarctic to the Arctic and all points in between. With the other editors of this book, he has led the efforts to increase the overall knowledge of soils and their relationship to global climate change. He has collaborated with Dr. Rattan Lal, Dr. Ronald Follett, and others to produce 12 books related to the role of soils in global climate change.

# Contents

## Section III    Other Forms of Degradation

## Section IV    Policy and Conservation Programs

# SECTION I

# Basic Processes

# CHAPTER 1

# Soil Degradation

## 1.1 SOIL

Soil is a three-dimensional body that forms the uppermost layer of the earth's crust. This layer supports all terrestrial life, filters and purifies water, biodegrades pollutants, and moderates gaseous exchange between terrestrial–aquatic ecosystems and the atmosphere. It is the most basic of all natural resources and the primary medium for food production. In addition to a source of industrial raw material and foundation for civil structures, it is also an archive of planetary history and the repository of germplasm or seedbank. The term soil resource is defined as beneficial assets and properties in relation to the use for which the soil is intended.

Sustainable management of soil depends on a thorough understanding of its attributes, the processes moderating its ecosystem services or functions of terrestrial importance, and transformations that occur through its interaction with the environment. Important attributes of soil include the following:

- It is nonrenewable over the human timescale of decades to centuries.
- It is unequally distributed over the landscape and among biomes/ecoregions.
- It is susceptible to misuse and mismanagement.

The term susceptible, when applied to soil, implies the possibility of adverse changes in soil properties and processes that lead to a reduction in the soil's ability to perform ecosystem functions. The products of these adverse changes to soils are collectively termed soil degradation. Soil in its natural state is in a dynamic equilibrium with its environment (Figure 1.1). It strongly interacts with the biosphere and is teeming with micro- and macrolife. The biotic activity alters soil properties, and soil properties in turn support specific life form(s) adapted to these changes in its attributes. Devoid of life, soil would cease to perform many ecosystem services. Important attributes emerging through interaction with the biosphere include the soil organic carbon (SOC) pool and its dynamics, elemental (e.g., C, N, P) cycling, consumption of oxygen ($O_2$), and production of greenhouse gases (GHGs) including carbon dioxide ($CO_2$), methane ($CH_4$), and nitrous oxide ($N_2O$). Soil's interaction with the

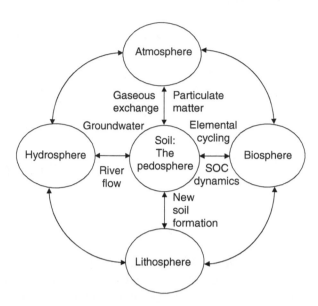

**Figure 1.1**    Soil is in a dynamic equilibrium with its environment.

atmosphere leads to gaseous exchange. The soil atmosphere contains higher concentrations of several trace gases and GHGs such as $CO_2$ and $N_2O$ than the atmosphere. In natural ecosystems (such as under the natural vegetation, well-drained conditions, etc.), soil is a sink for $CH_4$, oxidizing it into $CO_2$ and $H_2O$. Under managed ecosystems (monoculture, poorly drained conditions), however, soil can be a source of $CH_4$. Soil's interaction with the hydrosphere influences shallow groundwater, deep aquifers, springs, and stream flow. Soil is a major reservoir of fresh water on which all plant life depends. Changes over time to the lithosphere lead to new soil formation and geologic erosion that creates alluvial flood plains and loess deposits comprising some of the most fertile soils. In its natural state, soil is able to adapt/adjust to changes that occur on a geologic timescale. It is the rapid changes due to anthropogenic perturbations to which soil is unable to adjust. The natural changes occurring on a geologic timescale are caused by soil-forming factors and lead to soil formation.

## 1.2 SOIL DEGRADATION

Degradation means adverse changes in attributes leading to a reduced capacity to function. Thus, soil degradation means adverse changes in soil properties and processes over time. These adverse changes can be set in motion by disturbance of the dynamic equilibrium of soil with its environment (Figure 1.1) either by natural or anthropogenic (human) perturbations. Natural perturbations (e.g., shift in vegetation, glaciation, climate change) are often slow, allowing soil to adjust or adapt to the new conditions. Some natural perturbations can be rapid with drastic changes

Table 1.1   Types of Soil Degradation

| Type | Degradation Process |
|------|---------------------|
| Physical | Breakdown of soil structure |
| | Crusting and surface sealing |
| | Compaction, surface, and subsoil |
| | Reduction in water infiltration capacity |
| | Increase in runoff rate and amount |
| | Inundation, waterlogging, and anaerobiosis |
| | Accelerated erosion by water and wind |
| | Desertification |
| Chemical | Leaching of bases |
| | Acidification |
| | Elemental imbalance with excess of Al, Mn, Fe |
| | Salinization, alkalinization |
| | Nutrient depletion |
| | Contamination, pollution |
| Biological | Depletion of the soil organic carbon pool |
| | Decline in soil biodiversity |
| | Increase in soil-borne pathogens |

(e.g., tectonic activity, volcanic eruption). In general, however, anthropogenic activities are rapid, disturb the delicate balance between soil and its environment, and lead to drastic alterations in soil properties and processes. It is these adverse alterations in soil properties and processes due to anthropogenic perturbations that end in soil degradation, which is a major concern of the modern era.

Soil degradation is defined as diminution of soil's potential or actual utility, and reduction in its ability to perform ecosystem functions. In other words, it is the decline in soil quality leading to a reduction in biomass productivity and environment (water and air) moderation capacity. Three types of soil degradation processes — physical, chemical, and biological — are shown in Table 1.1 (Lal et al., 1989).

## 1.2.1   Soil Physical Degradation

The physical degradation process affects soil's mass to volume relationship, air to water relationship, gaseous exchange between soil and the atmosphere, and resistance against disruptive forces of air and water. These processes affect the soil's physical, mechanical, rheological (flow), and hydrological properties. Principal physical degradative processes are defined as:

*Breakdown of soil structure* — a reduction in the proportion and strength/stability of aggregates

*Slaking* — the dispersion of aggregates on quick immersion in water

*Crusting or surface scaling* — formation of a thin crust on the soil surface characterized by high strength and low permeability to water and air

*Densification* — an increase in soil bulk density leading to reduction in total porosity and macroporosity

*Anaerobiosis* — a decrease in aeration porosity to less than 10% by volume resulting in a lack of oxygen ($O_2$) in plant roots

**Plate 1.1**   Sheet erosion on a newly planted cropland.

**Plate 1.2**   Rill erosion on a freshly plowed cropland.

*Erosion* — the detachment, breakdown, transport, and redistribution of soil particles by
forces of water (rain, concentrated flow, stream, glaciers), wind, or gravity. The water
erosion may be sheet erosion (Plate 1.1), rill erosion (Plate 1.2), erosion by concentrated flow (Plate 1.3), or terrain deformation by severe gully erosion (Plate 1.4).

*Desertification* — soil degradation (by accelerated erosion by water, wind, and other
processes) in arid and semiarid climates leading to encroachment of desert-like
conditions (Plate 1.5A and Plate 1.5B)

**Plate 1.3**  Water erosion by concentrated flow on a plowed cropland.

**Plate 1.4**  Terrain deformation by mining and the attendant severe gully erosion.

**Plate 1.5A**   Decline in vegetation cover is indicative of desertification.

**Plate 1.5B**   Desertification implies spread of desert-like conditions by anthropogenic activities.

## 1.2.2   Soil Chemical Degradation

Chemical degradation processes refer to adverse changes in soil reaction or pH, reduction in reserves and availability of plant nutrients, the ability to inactivate toxic compounds and reduce excessive buildup of salts in the root zone. Principal chemical processes are as follows:

*Acidification* — a decline in soil pH caused by leaching of bases or addition of acid-producing fertilizers

*Nutrient depletion* — the removal of essential plant nutrients (e.g., N, P, K, Ca, Mg, Zn) by harvesting plants or excessive leaching without replenishment with inorganic or organic amendments

**Plate 1.6**   Secondary salinization is caused by poor soil drainage and excessive use of poor quality water.

*Toxification* — excessive buildup of some elements (e.g., Al, Mn, Fe) to a level that is toxic to plants

*Salinization* — excessive buildup of soluble salts in the root zone such that electrical conductivity of the saturated base exceeds 4 ds /cm (Plate 1.6)

*Alkalinization* or sodication — the predominance of sodic salts (Na ion) in the root zone leading to a sodium absorption ratio (SAR) of 15 and soil pH of > 8.5

*Pollution/contamination* — the application of industrial, mine waste, and urban pollutants to soil

## 1.2.3   Soil Biological Degradation

Changes in soil biological processes can increase soil degradation and adversely affect the quantity and quality of the SOC pool activity and species diversity of soil biota, and increase in relative proportion of soilborne pathogens. Characterizing such changes are:

*Soil organic carbon depletion* — a reduction in total and microbial biomass carbon, changes in turnover rate of the SOC pool

*Decline in soil biodiversity* — a reduction in activity and specific diversity of favorable organisms (e.g., earthworms) and shift in species composition

Soil physical, chemical, and biological degradative processes interact with one another, and exacerbate the adverse impacts on biomass production and the environment (Figure 1.2). In addition to decreasing agronomic crops, pasture, and forestry productivity, soil degradation can increase pollution of water bodies and decrease air quality.

**Table 1.2   Factors of Soil Degradation**

| Factor | Degradation Process |
|---|---|
| Soil | Erosion, compaction, crusting, anaerobiosis, nutrient depletion |
| Parent material | Salinization, alkalinization, nutrient reserve and availability, hard-setting |
| Climate | Desertification, acidification, leaching, SOC depletion, toxification |
| Terrain | Anaerobiosis, erosion, SOC depletion, nutrient imbalance |
| Vegetation | Type and quality of biomass returned to soil, soil biodiversity, nutrient cycling, water and energy balance |

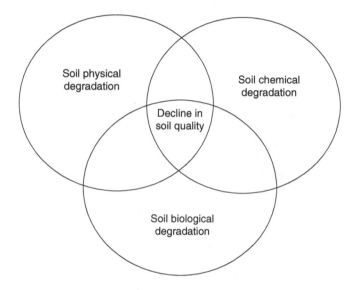

**Figure 1.2**    Interaction among soil degradative processes leads to decline in soil quality.

## 1.3 FACTORS OF SOIL DEGRADATION

Factors of soil degradation are biophysical environments that determine the type (e.g., physical, chemical, or biological) of degradation process (Table 1.2; Lal, 1997). Factors of soil degradation are analogous to Jenny's factors of soil formation (Jenny, 1941) (Equation 1.1):

$$S_d = f(S, C, T, V, M)_t \qquad\qquad (1.1)$$

where $S_d$ is soil degradation, S is soil properties, C is climatic parameters, T is terrain, V is vegetation, M is management, and t is time. Important factors of soil degradation include the following:

*Soil properties:* Inherent characteristics of soil that affect the type of soil degradation. These characteristics include texture, clay minerals, structure, horizonation, etc. The nature of parent material (type of rock) is an important factor of soil degradation.

*Climate*: Important climate parameters that affect the type of soil degradation are precipitation, temperature, seasonality, and evapotranspiration. Wind direction and velocity affect the rate and severity of wind erosion, just as rainfall amount and intensity determine the magnitude of water erosion. Leaching and acidification depend on the amount and distribution of rainfall.

*Terrain*: Important parameters of terrain that affect the type of soil degradation include slope gradient and slope length. The slope aspect (whether north or south facing) determines soil temperature, vegetation cover, weathering, and soil-water regime. Slope shape refers to the regularity of the slope. The shape (e.g., regular, convex, concave, or complex) affects the rate of erosion and deposition. The terrain also influences the drainage density and hydrological characteristics.

*Vegetation*: Important parameters of vegetation that affect the type of soil degradation include the percent ground cover and canopy height which affect soil erosion, species composition and succession which affect elemental cycling and SOC dynamics, and strata that affect microclimate and decomposition processes.

## 1.4 CAUSES OF SOIL DEGRADATION

Soil degradation is a biophysical process, driven by socioeconomic and political causes (Figure 1.3; Lal, 1997). Whereas the type of soil degradation is influenced by factors of soil degradation, the rate and severity of degradation processes are determined by causes of soil degradation. Causes of soil degradation may be specific to the type of soil degradation process (Table 1.3).

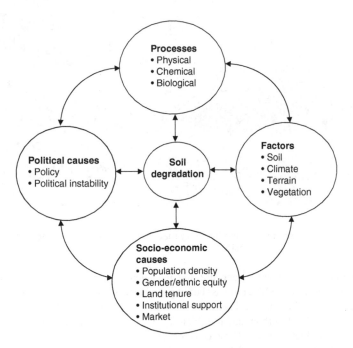

**Figure 1.3**   Interaction among processes, factors, and causes of soil degradation.

Table 1.3 Causes of Soil Degradation

| Type | Causes |
|------|--------|
| Physical | Deforestation |
| | Biomass burning |
| | Denudation |
| | Tillage up and down the slope |
| | Excessive animal, human, and vehicular traffic |
| | Uncontrolled grazing |
| | Monoculture |
| Chemical | Excessive irrigation with poor quality water |
| | Lack of adequate drainage |
| | No, little, or excessive use of inorganic fertilizers |
| | Land application of industrial/urban wastes |
| Biological | Removal and/or burning of residues |
| | No or little use of biosolids (e.g., manure, mulch) |
| | Monoculture without growing cover crops in the rotation cycle |
| | Excessive tillage |

**Plate 1.7** Deforestation and logging by heavy machinery cause soil compaction and make the soil susceptible to erosion.

*Biophysical causes* involve land use and management. Important among these are deforestation (Plate 1.7), biomass burning, drainage and irrigation, cropping/farming systems, fertilizer use, tillage methods (Plate 1.8), grazing intensity (Plate 1.9), logging methods, urban land use (Plate 1.10A and Plate 1.10B), and mining (Plate 1.11A and Plate 1.11B). These causes accentuate compaction, erosion, water run-off, anaerobiosis (Plate 1.12A to Plate 1.12D), nutrient depletion, reduction in SOC pool, etc.

**Plate 1.8** Tillage decreases soil structure and exacerbates the degradative processes.

**Plate 1.9** Excessive grazing at a high stocking rate leads to degradation of grazing lands.

*Socioeconomic causes* include demographic parameters (such as population density, education, gender/ethnic equity), land tenure, institutional support, and access to market.

*Political causes* include political instability, which is an important cause of soil degradation. In contrast, policy incentives that enhance investments in soil would reduce the rate and extent of soil degradation.

**Plate 1.10A**   Conversion of agricultural land to nonagricultural uses is occurring in periurban areas.

**Plate 1.10B**   Urban sprawl is encroaching upon some of the U.S. cropland.

## 1.5 LAND VS. SOIL DEGRADATION

The term land is broad based and implies an ecosystem comprising soil, micro- and mesoclimate, vegetation, terrain, and hydrology. The land includes indigenous and exotic plants and animals, soil biota, streams and ground water, and microclimate (Conacher and Conacher, 1995). The United Nations Environmental Program (UNEP) defines land as a concept that "includes soil and local water resources, land

**Plate 1.11A**   Mining causes drastic soil disturbance with adverse impacts on the environment.

**Plate 1.11B**   Mining causes drastic soil disturbance.

surface and vegetation or crops that may be affected by one or a combination of processes acting on it" (UNEP, 1993). Therefore, the term land degradation implies decline in quality and capacity of the life-supporting processes and environment moderation capacity of the ecosystem through adverse changes in all components including soil, vegetation, climate, hydrology, and terrain resulting from human actions. Conacher and Conacher (1995) defined land degradation as "alterations to all aspects of the natural (or biophysical) environment by human actions, to the

**Plate 1.12A**    Poor drainage is caused by soils with slow infiltration rate.

**Plate 1.12B**    Anaerobiosis is also caused by soil compaction, especially in the head lands.

detriment of vegetation, soils, landforms and water (surface and subsurface, terrestrial and marine) and ecosystems." Johnson and Lewis (1995) defined land degradation as "the substantial decrease in either or both of an area's biological productivity or usefulness due to human interference." Similar to soil, land is also in a dynamic equilibrium with its environment. An alteration of one component (e.g., vegetation) can lead to drastic changes in others (e.g., soil, hydrology, microclimate). The system can recover if the perturbation is slight and causes only temporary changes in the ecosystem's functions. An example of a temporary change is nutrient

**Plate 1.12C**   Inundation in depressional sites occurs following heavy or frequent rains.

**Plate 1.12D**   Some croplands suffer severe yield losses due to inundation and waterlogging.

depletion, which can be easily remedied by addition of nutrients. The system may undergo irreversible changes and not recover to its original state if the perturbation is drastic (both in duration and intensity). An example of drastic perturbation is large-scale deforestation leading to severe gullying of a steep terrain.

It is, therefore, important to make a clear distinction between land degradation and soil degradation. Using these two terms interchangeably has caused confusion, misinterpretations, and poor planning.

## 1.6 SOIL DEGRADATION VS. SOIL FORMATION

The disturbances or perturbation of the dynamic equilibrium leads to changes in soil properties and processes. The change may be positive leading to aggradation or improvements in soil productivity and environment moderation capacity, or negative leading to degradation or decline in soil productivity and environment moderation capacity. If the cause of a disturbance is natural, the resultant change is a natural soil-forming process. On the other hand, if the cause of disturbance is human induced or anthropogenic, the attendant decline in soil quality is soil degradation. Therefore, soil degradation is the loss of actual or potential productivity and capacity to perform ecosystem functions due to anthropogenic causes. In other words, soil formation is a natural process and soil degradation is a human-induced process.

## 1.7 SOIL QUALITY

Soil quality refers to the capacity of the soil to perform specific functions (Lal, 1993; 1997; 1998; 1999; Doran and Parkin, 1994; Doran et al., 1996; Carter et al., 1997; Larson and Pierce, 1991; Bezdicek et al., 1996; Karlen et al., 1997; Papendick and Parr, 1992; Parr et al., 1992). Important functions of soil include agronomic/biomass productivity, purification of water, biodegradation of pollutants, moderating emission of GHGs from soil to the atmosphere, foundation for civil/urban structures, etc. Soil quality is an indication of "the state of the soil system" (Jenny, 1980), and is determined by the interrelationships among soil properties, soil type, land use and management. Thus, soil quality ($S_q$) for agronomic/forestry use is a function of several attributes as is shown in Equation 1.2 (Lal, 1993):

$$S_q = f(P_i, S_c, R_d, e_d, N_c, B_d)_t \qquad (1.2)$$

where $P_i$ is productivity, $S_c$ is an index of structural properties, $R_d$ is rooting depth, $e_d$ is charge density, $N_c$ is nutrient reserve, $B_d$ is a measure of soil biodiversity, and t is time. In this context, soil degradation ($S_d$) is decline in soil quality (Equation 1.3).

$$S_d = -dS_q/dt \qquad (1.3)$$

In other words, soil degradation implies a decline in soil's inherent capacity to produce economic goods and provide ecosystems services.

## 1.8 SOIL EXHAUSTION VS. SOIL DEGRADATION

Exhaustion refers to system fatigue due to overuse (Lal, 1997). Therefore, soil exhaustion implies a decline in productivity and the environment moderation capacity due to overuse or intensive use of soil. Because the decline is not due to misuse or mismanagement, soil productivity cannot be restored even with additional production effort such as use of fertilizer amendments. Soil exhaustion, however, is a temporary change that can be remedied through change in land use. Therefore, an important strategy would be to change the specific land use or grow a cover crop in the rotation cycle.

## 1.9 SOIL RESILIENCE

Resilience refers to the capacity of a system to absorb change without significantly altering the relationship between the relative importance and numbers of individuals and species of which the community is composed (Douglas and Lewis, 1995). Therefore, soil resilience refers to the ability of soil to resist or recover from an anthropogenic or natural perturbation (Barrow, 1991; Lal, 1997). Soil resilience ($S_r$) can be assessed through a mass balance of one or several interrelated characteristics (Lal, 1993; 1997) as shown in Equation 1.4:

$$S_r = S_a + \int_0^t (S_n - S_d + I_m)dt \qquad (1.4)$$

where $S_a$ is the initial or antecedent soil condition, $S_n$ is the rate of soil renewal, $S_d$ is the rate of soil degradation, $I_m$ is the management input and t is time. Assessment of $S_r$ using Equation 1.4 is applicable more to a specific soil property (e.g., SOC content, nutrient reserve) than to a soil system on the whole. There is also a close interdependence between soil quality and soil resilience. Resilient soils have a high soil quality and vice versa (Equation 1.4). The negative value of the right-hand side of Equation 1.5 implies soil degradation (Lal, 1997):

$$S_r = dS_q/dt \qquad (1.5)$$

## 1.10 SOIL DEGRADATION AND DESERTIFICATION

The term desertification implies the spread of desert-like conditions. It is an all-encompassing term and suggests the complexity of numerous processes involved in the increase of desert-like conditions. Important processes include drought, erosion by water and wind, and denudation of vegetation cover. UNEP (1977) defined desertification as "the diminution or destruction of the biological potential of land which can lead ultimately to desert-like conditions." The working definition has since been refined to imply "the land degradation in arid, semi-arid and dry sub-humid areas resulting from various factors including climatic variations and human activities" (UNEP, 1990; 1992; UNCED, 1992). Desertification may refer to a process or the ultimate stage of land degradation (Mainguet, 1991).

## 1.11 CONCLUSION

Soil degradation has challenged mankind since the dawn of civilization, but has been an important issue since the 1930s (Jacks and Whyte, 1939; Osborn, 1948; Carson, 1962; Commoner, 1972; Blakie and Brookfield, 1987; Barrow, 1991).

Soil degradation and its environmental consequences will remain an important issue during the 21st century. It is simply defined as the decline in soil quality caused through its misuse by humans. While natural factors may cause or even intensify the process, soil degradation as defined in this book refers to that resulting explicitly from human actions. There are three principal types of soil degradative processes: physical, chemical, and biological. Together, interactive effects of these three processes lead to decline in soil quality over time. Soil degradation may be temporary and reversed through change in land use and management, or it may be permanent or irreversible. Soil quality of slightly or temporarily degraded soils can be restored through change in land use and management. The soil's ability to restore or recover its quality, in terms of biomass productivity and environment moderation capacity, is termed "soil resilience."

While recognizing that this is a global problem, the objective of this volume is to provide a state-of-the-knowledge compendium on the extent and severity of soil degradation by different processes in the U.S. The objective is to:

- Compile the available information on soil degradation.
- Identify temporal changes in soil degradation during the last three decades of the 20th century.
- Provide a framework to assess if and where soil degradation poses a threat to productivity of agriculture and forestry, and to the environment.
- Identify the need for any policy options that would lead to sustainable use of soil resources of the U.S.

The book specifically focuses on soil degradation of agricultural areas including cropland, pasture land, and range land. Additional information has also been included on desertification and loss of wetlands in the U.S. The book is based on the data from National Resources Inventory reports compiled by the Natural Resources Conservation Service (formerly Soil Conservation Service of the U.S. Department of Agriculture).

## REFERENCES

Barrow, C.J. *Land Degradation: Development and Breakdown of Terrestrial Environments.* Cambridge University Press, U.K., 1991, 235 pp.

Bezdicek, D.F., R.I. Papendick, and R. Lal. Importance of soil quality to health and sustainable land management. In *Methods for Assessing Soil Quality,* Special Publication No. 49. J.W. Doran and A.J. Jones (Eds.), Soil Science Society of America, Madison, WI, 1996, 1–7.

Blakie, P. and H. Brookfield. *Land Degradation and Society.* Methuen, London, 1987.

Carson, R.H. *Silent Spring.* Houghton Mifflin, Boston, 1962.

Carter, M.R., E.G. Gregorich, D.W. Anderson, J.W. Doran, H.H. Janzen, and F.J. Pierce. Concepts of soil quality and their significance. In *Soil Quality for Crop Production and Ecosystem Health.* E.G. Gregorich and M.R. Carter (Eds.), Elsevier, Amsterdam, 1997, 1–19.

Commoner, B. *The Closing Circle: Confronting the Environmental Crisis.* Jonathan Cape, London, 1972.

Conacher, A. and J. Conacher. *Rural Land Degradation in Australia.* Oxford University Press, Oxford, U.K., 1995, 170 pp.

Doran, J.W. and T.B. Parkin. Defining and assessing soil quality. In *Defining Soil Quality for a Sustainable Environment*, Special Publication No. 35. J.W. Doran, D.C. Coleman, D.F. Bezedick, and B.A. Stewart (Eds.), Soil Science Society of America, Madison, WI, 1994, 3–21.

Doran, J.W., M. Sarrantonio, and M.A. Lieberg. Soil health and sustainability. *Adv. Agron.* 56, 1–54, 1996.

Douglas, L.J. and L.A. Lewis. *Land Degradation: Creation and Destruction.* Blackwell, Oxford, U.K., 1995, 335 pp.

Jacks, G.V. and R.O. Whyte. *The Rape of the Earth: A World Survey of Soil Erosion.* Faber and Faber, London, 1939.

Jenny, H. *Factors of Soil Formation.* McGraw Hill, New York, 1941, 281 pp.

Jenny, H. *The Soil Resources.* Springer-Verlag, New York, 1980.

Johnson, D. and L. Lewis. *Land Degradation: Creation and Destruction.* Blackwell Scientific Publishers, Oxford, U.K., 1995, 335 pp.

Karlen, D.L., M.J. Mausbach, J.W. Doran, R.G. Cline, R.F. Harris, and G.E. Schuman. Soil quality: A concept, definition and framework for evaluation. *Soil Sci. Soc. Am. J.* 61, 4–10, 1997.

Lal, R. Tillage effects on soil degradation, soil resilience, soil quality and sustainability. *Soil Tillage Res.* 27, 1–8, 1993.

Lal, R. Degradation and resilience of soils. *Philos. Trans. R. Soc. (Biology) London* 352, 997–1010, 1997.

Lal, R. *Soil Quality and Agricultural Sustainability.* Ann Arbor Press, Chelsea, MI, 1998.

Lal, R. *Soil Quality and Soil Erosion.* CRC Press, Boca Raton, FL, 1999.

Lal, R., G.F. Hall, and F.P. Miller. Soil degradation: basic processes. *Land Degradation Rehabil.* 1, 51–69, 1989.

Larson, W.E. and F.J. Pierce. Conservation and enhancement of soil quality. In *Evaluation for Sustainable Land Management in the Developing World*, IBSRAM, Bangkok, Thailand, 1991.

Mainguet, M. *Desertification: Natural Background and Human Mismanagement.* Springer-Verlag, Berlin, 1991.

Osborn, F. *Our Plundered Planet.* Faber and Faber, London, 1948.

Papendick, R.I. and J.F. Parr. Soil quality—the key to a sustainable agriculture. *Am. J. Alternative Agric.* 7, 2–3, 1992.

Parr, J.F., R.I. Papendick, S.B. Hornick, and R.E. Meyer. Soil quality: Attributes and relationship to alternative and sustainable agriculture. *Am. J. Alternative Agric.* 7, 5–11, 1992.

UNCED. Earth summit Agenda 21: Program of Action for Sustainable Development. UNEP, New York, 1992.

UNEP. World map of desertification. UN Conference on Desertification. August 29–September 9, 1977. Document A/Conf. 74/2, Nairobi, Kenya.

UNEP. Desertification revisited. UNEP/DC/PAC, Nairobi, Kenya, 1990.

UNEP. World Atlas of Desertification. Edward Arnold, Sevenoaks, UNEP, Nairobi, Kenya, 1992.

UNEP. World map of present day landscapes: an explanatory note. United Nations Environment Program and Moscow State Univ., Moscow, 1993.

CHAPTER 2

# Global Extent of Soil Degradation and Desertification

## 2.1 BACKGROUND

Soil degradation has plagued mankind for thousands of years. Archaeological evidence shows that soil degradation was responsible for extinction of many ancient civilizations (Lowdermilk, 1953; Olson, 1981), yet the available statistics on the extent and severity of soil degradation are not reliable. Some statistics in the literature indicate that the world is indeed running out of high-quality soils. If that is correct, then the question remains why urgency is lacking on the part of planners to do something about this important issue.

Seeking an objective answer to this question requires separating emotional rhetoric from facts, obtaining reliable/credible statistics on the extent and severity of soil degradation by different processes, establishing the cause-effect relationship, and identifying land use and soil/vegetation/water management strategies that would reverse the degradative trends. Indeed, if the available statistics on soil degradation are correct, there is a definite cause for concern, and this is the greatest challenge facing the future of the planet. Even if the available statistics are nearly correct and their consequences as drastic as often portrayed, it is indeed perplexing that urgency is lacking on the part of the global community and policymakers to do something about it. Is it possible that the scientific community has overdramatized the issue? If so, there is a credibility problem. If the statistics are correct, there is a need to develop a global coordinated effort to address this severe and menacing challenge facing all of us.

## 2.2 DATABASE, METHODOLOGY, AND EXTRAPOLATION

Soil degradation is widely recognized as a serious problem. Unknown, however, are the extent, severity, and rate of soil degradation by different processes in principal biomes/ecoregions of the world and their economic and environmental consequences

23

Table 2.1    Three Different Estimates of Land Area Affected by Desertification

| Region | FAO/UNESCO/UNEP (1980) | Dregne (1983) | Mabbutt (1984) |
|---|---|---|---|
| | $10^6$ km$^2$ | | |
| Africa | 13.7 | 10.7 | 7.4 |
| Asia | 10.4 | 12.0 | 7.5 |
| Australia | 5.7 | 3.1 | 1.1 |
| Europe | 0.2 | 0.2 | 0.3 |
| North and Central America | 4.3 | 3.2 | 2.1 |
| South America | 3.3 | 3.3 | 1.6 |
| Total | 37.6 | 32.5 | 20.0 |

Table 2.2    Estimates of Soil Erosion by Water in India and Pakistan

| Country | National Estimate | GLASOD Estimate |
|---|---|---|
| | $10^6$ ha | |
| India[a] | 87–111 | 32.8 |
| Pakistan[b] | 11.2 | 7.2 |

[a] RAPA (1992); Sehgal, J. and Abrol, I. P. (1992).
[b] Mian and Javed (1989); RAPA (1990).
Source: From FAO, World Soil Resour. Rep. 78, 1994.

(Lal, 1994). Despite its significance, the available information on soil degradation is often based on reconnaissance surveys, public opinion, extrapolations based on sketchy data, and casual observations by interested travelers.

The quality of available data on the global extent of soil degradation is extremely uneven. Consequently, two estimates about the same degradation process may differ by an order of magnitude (Lal, 1994). One of the problems is the lack of quantitative criteria in defining a specific soil degradative process. An example highlighting such a problem is presented in Table 2.1 which shows global estimates of desertification. The criteria used are different for each group and produce different estimates. Hence, large differences in the estimates of the degraded/desertified area are reported.

In addition to the criteria, there is also a problem with the methodology. Data obtained by diverse and unstandardized methods can lead to gross errors. Estimates of soil erosion by water in India and Pakistan in Table 2.2 differ widely due to differences in methods used to obtain these data. It is important to realize that unreliable data are worse than no data because the unreliable and erroneous information can lead to misleading interpretations, and expensive and counterproductive measures. The Food and Agricultural Organization (FAO) (FAO, 1979; FAO/UNED/ UNESCO, 1980) developed a provisional methodology for assessment of soil degradation at the regional scale. This was done without a strong and credible database. Thus, it is difficult to validate the information and use it toward planning for sustainable management of soil resources.

The literature (Pimentel et al., 1995; Dregne, 1998) on soil degradation is also replete with gross extrapolations based on scanty data, often outside the ecoregions from which data were obtained. Such extrapolations are not valid, and are grossly misleading with regard to making management decisions. There are also problems with scaling. Soil degradation processes (e.g., erosion) are scale dependent. Scaling problems can lead to severe discrepancies in estimates. Lal (1994) reported that global denudation rates ranged from 0.04 to 0.30 mm/yr, a difference of 7.5 times, depending on the scaling procedure used.

Because of its global significance, it is important to develop reliable and credible databases on soil degradation. Subjective and qualitative information gathered by unstandardized methods must be replaced by objective and quantitative data obtained by standardized procedures that produce repeatable information verifiable by ground truth measurements. There is much literature on soil degradation, but there are few reports containing quantitative and credible data.

## 2.3 GLASOD METHODOLOGY

The Global Assessment of Soil Degradation (GLASOD) project was organized by the International Soil Reference and Information Center (ISRIC) located in Wageningen, Holland (Oldeman, 1988; Oldeman et al., 1991; Oldeman, 1994; Oldeman and van Lynden, 1998). The project was funded by the United Nations Environment Program (UNEP) and the FAO of the United Nations. The GLASOD methodology involved a "survey of expert opinion" on the status of the human-induced soil degradation, and was conducted in close consultation with national soil scientists. The information was collected for 21 regions and by 250 scientists from around the world according to general guidelines outlined by ISRIC (Oldeman, 1988). Using the FAO/UNEP/UNESCO (1980) definition of soil degradation ("the human-induced phenomena which lowers the current and/or future capacity of the soil to support human life"), the type (process) and degree (severity) of soil degradation were defined. This methodology assessed the "present" state of soil degradation without considering either the rate or the potential (future) hazards. Based on these methodologies, the global extent of soil degradation by different processes is outlined in Table 2.3, severity of soil degradation in Table 2.4, and the causes of

Table 2.3  Global Extent of Soil Degradation by Different Processes

| Type of Degradation Process | Land Area Affected ($10^6$ ha) |
|---|---|
| Water erosion | 1094 |
| Wind erosion | 548 |
| Chemical degradation | 472 |
| Physical degradation | 82 |
| Total | 1964 |

Note: The total is not sum of all processes because two or more processes may affect the same land area.

Source: Modified from Oldeman, L. R., in Soil Resilience and Sustainable Land Use, CAB International, Wallingford, U.K., 1994 and Oldeman, L. R. and van Lynden, G. W. J., in Methods for Assessment of Soil Degradation, CRC Press, Boca Raton, FL, 1998, 423–440.

Table 2.4    Severity of Soil Degradation by
             Different Processes at Global Scale

| Severity | Land Area Affected (10⁶ ha) |
|---|---|
| Light | 749 |
| Moderate | 910 |
| Severe and extreme | 305 |
| Total | 1964 |

*Source*: Modified from Oldeman, L. R., in *Soil Resilience
and Sustainable Land Use,* CAB International, Wallingford, U.K., 1994 and Oldeman, L. R. and van Lynden,
G. W. J., in *Methods for Assessment of Soil Degradation*,
CRC Press, Boca Raton, FL, 1998, 423–440.

Table 2.5    Causes of Global Soil Degradation

| Causes | Land Area Affected (10⁶ ha) |
|---|---|
| Deforestation | 579 |
| Overexploitation | 133 |
| Overgrazing | 679 |
| Agricultural activities | 552 |
| Industrial activities | 23 |
| Total | 1966 |

*Note:*   The total land area affected (1966 Mha) is
          slightly different than that listed in Table 2.3
          (1964 Mha) due to error in rounding numbers.

*Source*: Modified from Oldeman, L. R., in *Soil Resilience and Sustainable Land Use,* CAB International,
Wallingford, U.K., 1994.

soil degradation in Table 2.5. The total land area affected by all types and severity of
soil degradation is about 2 billion hectares out of the total land area of 13 billion hectares.

Useful as these data are, there are several limitations that need to be considered
while interpreting the information. The information is based on expert judgement,
and is thus subjective. Further, the data set is qualitative, available only at the
continental scale, and not related to productivity for different land uses or soil
management options. These inherent methodological flaws can be removed only by
development of a methodology founded on basic processes at the "pedon" or
"soilscape" level and then scaling up using Geographic Information Systems (GIS).

## 2.4   DESERTIFICATION ASSESSMENT

The first global assessment of desertification was published by UNEP (1977),
and later revised in 1992 (UNEP, 1992). This assessment was based on a country-
by-country report started in 1975 (Dregne et al., 1991). Later, this assessment was
combined with the data obtained by GLASOD methodology. These maps were
prepared at the scale of 1:25,000,000 and at a reduced scale of 1:150,000,000. Thus,
mapping unit boundaries are vague at best. Assessment of desertification in relation
to land use has been made by Dregne and Chou (1992). Most of this information,

**Table 2.6  Global Assessment of Desertification**

| UNEP (1991) | | Oldeman and van Lynden (1998) | |
|---|---|---|---|
| Land Type | Area Affected (10⁶ km²) | Type of Soil Degradation | Land Area Affected (10⁶ km²) |
| Irrigated land | 0.43 | Water erosion | 4.78 |
| Rainfed cropland | 2.16 | Wind erosion | 5.13 |
| Range land (soil and vegetation degradation) | 7.57 | Chemical degradation | 1.11 |
| Subtotal | 10.16 | Physical degradation | 0.35 |
| Range land (vegetation degradation only) | 25.76 | | |
| Total | 35.92 | Total | 11.37 |

**Table 2.7  UNEP Estimates of the Land Area Susceptible to Desertification**

| Parameter | UNEP (1977) | UNEP (1984) | UNEP (1992) |
|---|---|---|---|
| Total land area (Mha) | 5550 | 4409 | 5160 |
| Percent of area susceptible to desertification | 71.5 | 78.9 | 69.6 |
| Land area susceptible to desertification (Mha) | 3970 | 3475 | 3592 |

**Table 2.8  Estimates of the Annual Rate of Desertification**

| Land Use | Total Land Area (Mha) | Desertification Rate | |
|---|---|---|---|
| | | Area (Mha/yr) | % of Total Area/Yr |
| Irrigated land | 131 | 0.125 | 0.095 |
| Rangeland | 3700 | 3.200 | 0.086 |
| Rainfed cropland | 570 | 2.500 | 0.439 |
| Total | 4401* | 5.825 | 0.132 |

*Source*: Calculated from Mainguet, M., *Desertification: Natural Background and Human Mismanagement*, Springer-Verlag, Berlin and UNEP (1991), *Environmental Data Report*, 3rd ed., Basil Blackwell, Oxford, 1991.

especially in relation to grazing land, is based on "informed opinion" (Dregne, 1989). The data on extent of desertification in relation to land use shown in Table 2.6 indicate the extreme susceptibility of rangeland. Of the 1.0 to 1.1 billion hectares of desertified soil (excluding vegetation), 0.99 billion hectares is due to water and wind erosion. Therefore, physical degradation is the principal type of the desertification process driven by inappropriate land use and poor soil management. UNEP assessments of the land area susceptible to desertification are shown by the data in Table 2.7. Total land area susceptible to desertification was estimated at 3970 million hectares (Mha) in 1977, 3475 Mha in 1984, and 3592 Mha in 1992. The data in Table 2.8 show the annual rate of desertification, which is about 5.8 Mha/yr. Most of these data are based on qualitative and subjective methodology.

The methodology and statistics used in desertification reports have limitations similar to those of the GLASOD assessment of soil degradation. The data and maps are vague and show nothing about where desertification problems exist and which are the critical areas (Dregne, 1998). A major problem with these statistics is the credibility of the data, because the maps and tables are based on expert opinion, and "anyone can pick a number out of the air and claim that it represents actual conditions." Another disadvantage is the scale. Even at 1:5,000,000 scale, the problem spots of up to 1000 ha are difficult to pinpoint and precisely locate. Consequently, extrapolations from a map at a scale of 1:25,000,000 or 1:150,000,000 can lead to gross errors.

## 2.5  THE NEED FOR A STRONG AND CREDIBLE DATABASE

The presently available statistics on the regional and global extents of soil degradation and desertification are vague, qualitative, subjective, and based on so-called informed opinion surveys. There are strong reservations about the methodology and its reproducibility, quality, accuracy, and reliability. These data cannot be used for land use planning and identification of conservation/restoration policies. There are also problems with using these data for assessing the impact of soil degradation and desertification on biomass/agronomic productivity, quality (eutrophication, contamination, pollution) of natural waters, and emission of greenhouse gases (GHGs) from soil/ecosystems into the atmosphere.

For land use planning and implementation of soil restoration strategies, it is important to develop a credible database at the national level. The data must be based on basic principles of soil science and reflect the types of soil degradation (e.g., erosion, salinization, waterlogging) so that the cause-effect relationship can be established, the extent and severity of degradation can be linked to the land use and management, effects on productivity and environment quality can be assessed, the data credibility can be verified by ground truthing, and the methodology used produces reproducible results. The objective of data collection is to address specific issues related to soil degradation effects on productivity, water and air quality, and emission of greenhouse gases. Credible data are needed to identify strategies for soil restoration to alleviate specific constraints.

There is also a strong need for more basic research to generate data that are accurate and credible, and devoid of emotional rhetoric. The emphasis is on obtaining reliable data collated by standard and reproducible methodology.

## REFERENCES

Dregne, H.E. *Desertification of Arid Lands.* Harwood Academic Publishers, New York, 1983.
Dregne, H.E. Informed opinion: filling the soil erosion data gap. *J. Soil Water Cons.* 44, 303–305, 1989.
Dregne, H.E. Desertification assessment. In *Methods for Assessment of Soil Degradation.* R. Lal, W.H. Blum, C. Valentine, and B.A. Stewart (Eds.), CRC Press, Boca Raton, FL, 1998, pp. 441–457.
Dregne, H.E., M. Kassas, and B. Rozanov. A new assessment of the world status of desertification. *Desertification Control Bull.* 20, 6–18, 1991.

Dregne, H.E. and N.T. Chou. Global desertification: dimensions and costs. In *Degradation and Restoration of Arid Lands*. H.E. Dregne (Ed.), Texas Tech. University, Lubbock, TX, 1992, pp. 249–281.

FAO. *A Provisional Methodology for Soil Degradation Assessment*. FAO, Rome, 1979, 73 pp.

FAO. Land Degradation in South Asia: Its Severity, Causes and Effects Upon the People. World Soil Resources Reports 78, FAO, Rome, 1994, 100 pp.

FAO/UNEP/UNESCO. Provisional map of present degradation rate and present state of soil: map scale of 1:5,000,000. FAO, Rome, 1980.

Lal, R. (Ed.) *Soil Erosion Research Methods,* 2nd ed. Soil Water Cons. Soc., Ankeny, IA, 1994, 340 pp.

Lowdermilk, W.C. Conquest of the land through 7000 years. USDA-SCS, Washington, D.C., 1953, 30 pp.

Mabbutt, J.A. A new global assessment of the status and trends of desertification. *Env. Cons.* 11, 100–113, 1984.

Mainguet, M. *Desertification: Natural Background and Human Mismanagement*. Springer-Verlag, Berlin, 1991.

Mian, A. and Y. Javed. The Soil Resources of Pakistan — Their Potential, Present State and Strategies for Conservation. Sector Paper of the National Conservation Strategy. Soil survey of Pakistan, Lahore, 1989, 53 pp.

Oldeman, L.R. (Ed.) Guidelines for general assessment of the status of human-induced soil degradation. ISRIC Working paper and preprint #88/4. Wageningen, The Netherlands, 1988.

Oldeman, L.R. Global extent of soil degradation. In *Soil Resilience and Sustainable Land Use*. D.J. Greenland and I. Szabolcs (Eds.), CAB International, Wallingford, U.K., 1994, pp. 99–118.

Oldeman, L.R., R.T.A. Hakkeling, and W.G. Sombroek. World map of the status of human-induced soil degradation. ISRIC, Wageningen, The Netherlands, 1990, 21 pp + 3 maps at scale of 1:5,000,000.

Oldeman, L.R. and G.W.J. Van Lynden. Revisiting the GLASOD methodology. In *Methods for Assessment of Soil Degradation*. R. Lal, W.H. Blum, C. Valentine, and B.A. Stewart (Eds.), CRC Press, Boca Raton, FL, 1998, pp. 423–440.

Olson, G.W. Archaeology: lessons on future soil use. *J. Soil Water Cons.* 36, 261–264, 1981.

Pimentel, D., C. Harvey, P. Resosudarmo, K. Sinclair, D. Kurz, M. McNair, S. Crist, L. Shpritz, L. Fitton, R. Saffouri, and R. Blair. Environmental and economic costs of soil erosion and conservation benefits. *Science* 267, 117–1133, 1995.

RAPA/FAO. Problem soils of Asia and the Pacific. RAPA Report 1990/6. FAO/RAPA, Bangkok, Thailand, 1990, 283 pp.

RAPA/FAO. Environmental issues in land and water development. FAO/RAPA, Bangkok, Thailand, 1992, 488 pp.

Sehgal, J. and I.P. Abrol. Land degradation status: India. *Desertification Bull.* 21, 24–31, 1992.

UNEP. United Nations Conference on Desertification (UNCD). World map of desertification. A/CONF. 74/2. UNEP, Nairobi, Kenya, 1977, 11 pp + map at 1:25,000,000.

UNEP. General assessment of the progress in the implementation of the Plan of Action to Combat Desertification 1977–84. UNE, Nairobi, Kenya, 1984.

UNEP. *Environmental Data Report*, 3rd ed. Basil Blackwell, Ltd., Oxford, 1991, 408 pp.

UNEP. *World Atlas of Desertification*. Edward Arnold, Hodder & Stoughton, Sevenoakes, U.K., 1992, 69 pp.

# CHAPTER 3

# Land Use in the U.S.

## 3.1 DEFINITIONS

Two widely used terms for denoting transformation of land from natural into managed ecosystems are: land use and land cover (Turner and Meyer, 1994). The term land use is primarily utilized by geographers, economists, anthropologists, and planners, and implies human use of the land for specific purposes. Land use is defined as "the human activity associated with a specific piece of land" (Lillesand and Kiefer, 1987) or "it is the purpose of human activity on the land" (NRCS, 1992). There are several classes of land use such as for growing crops, raising animals, establishing forests for timber, exploring for minerals and fossil fuels, establishing habitats and urban centers, preserving natural resources, and for recreation. From the knowledge of area under a specific land use at different times, it is possible to calculate "land use change" over that period or the rate of change of a specific land use. The land use change may involve shift from one category to another or to change in intensity of the same land use.

The term land cover refers to the physical state of the land, especially with regard to the surface area covered by vegetation, water, and earth materials. While the information on land use is of interest to social scientists, statistics on land cover is mainly used by natural scientists including soil scientists, agronomists, hydrologists, and climatologists. The land cover change may occur either by conversion or modification. An example of the conversion of land cover change is from cropland to pastureland (change from one class of land use to another). An example of modification of land cover is selective logging of a forest or switching to another species (from white pine to hard wood or poplar). Modification of land cover involves a change within the same class of land use.

Information on land use and land cover is useful for planning, assessing the impact on soil degradation, determining rates of desertification, changes in water quality and components of the hydrologic cycle, and possible shifts in the micro- and mesoclimate. Both land use and land cover are strongly influenced by human activities, and change drastically with the demographic pressure.

## 3.2 LAND USE CLASSES

There are several classes or types of land use. Predominant land use classes are briefly described below.

### 3.2.1  Forestland

Forestland is a perennial land use class. It is defined as "an ecological system with a minimum crown coverage of the land surface of more than 10%" (Williams, 1994). Forestlands are used for obtaining forest products, fuel wood, and products for industrial use. There are three categories of forestlands: natural forest including preserves, managed forest with selective cutting and plantation, or manmade forests. The natural forest may be closed or open (FAO, 1990). A closed forest is a natural, tree-dominated vegetation where trees cover a high proportion of the ground and grass does not form a continuous layer on the forest floor. In contrast, an open forest is a forest in which trees are interspersed with grazing land. The open forest is also sometimes called wooded land.

### 3.2.2  Grazing Land

This refers to the land on which animals are grazed. The term grazing implies feeding on growing grass or herbage. There are two classes of grazing land: rangeland and pastureland.

Rangeland is defined as "land on which the historic climax plant community or the potential vegetative cover is principally native grasses, grass-like plants, forbs or shrubs suitable for grazing and browsing" (NRCS, 1997). In comparison, pastureland implies "land used primarily for production of introduced or native forage plants for livestock grazing" (NRCS, 1992). In contrast to rangeland, pasturelands usually contain introduced (improved species) forage plant species to maximize land resources value related to forage production (Sobecki et al., 2001).

Although not specifically used by NCRS resources inventory, "grassland" is another land use type commonly referred to in the literature. Grasslands are "landscapes that have a ground story in which grasses are the dominant vegetation life form" (Graetz, 1994). Considering vegetation attributes as a principal criterion, the grassland class may also include the sparse woodlands with grassy understories, the savannas, and the very sparsely shrubbed grasslands of arid regions.

### 3.2.3  Cropland

Cropland is "land used for the production of adapted crops like corn, soybean, small grains, forages, and horticultural crops" (Heimlich and Daugherty, 1991). Thus, it is a human-made ecosystem. The land may be used continuously for these crops or the crops may be grown over a period of years in rotation with grasses and pastures. There are several types of cropland. Cultivated cropland includes land on which crops are grown and harvested, compared to noncultivated cropland on which crops are not grown during a specific period. Cultivated summer fallow is the land on which crops are not grown during the fallow period. Idle cropland includes land in which soil improvement crops or natural fallow (e.g., cover crops) are grown.

In terms of land use, U.S. cropland is of five distinct types: (1) harvested cropland consisting of land used continuously for growing crops and from which crops are harvested; (2) cropland that was sown to crops but on which crops failed and were not harvested; (3) cropland maintained as fallow and cultivated to control weeds and conserve water for the following crop; (4) idle cropland, which includes natural vegetation cover and soil improvement crops, or land left completely idle or without economic crops because of physical or economic reasons; and (5) cropland used for pasture grown in rotation with crops (Heimlich and Daugherty, 1991).

## 3.2.4 CRP Land

The land under the Conservation Reserve Program (CRP) was taken out of cultivation because it was highly erodible land (HEL). HEL is determined to have an inherent erosion potential of over eight times its soil loss tolerance (T) level (Uri, 2000). The determination of HEL is made by calculating the erodibility index (EI) for both water and wind erosion. The EI for water erosion is computed as shown in Equation 3.1:

$$EI = R*K*L*S/T \qquad\qquad (3.1)$$

where R is the rainfall or runoff factor, K is the soil erodibility factor, S is the slope steepness factor, and T is the soil loss tolerance. The EI for wind erosion is computed as shown in Equation 3.2:

$$EI = C*I/T \qquad\qquad (3.2)$$

where I is soil erodibility factor and C is the climatic factor based on wind velocity and surface soil moisture content.

As a provision of several Farm Bills, the HEL must be converted to a permanent vegetative cover for a 10-year period. The 1996 Farm Bill made a special provision to protect HEL. It stipulates that those HELs that best management practices (BMPs) cannot protect can be targeted for permanent land retirement. The principal objectives of CRP are to reduce transport of sediment and pollutants into natural waters (Dukes, 1996). Another important outcome of the CRP is carbon (C) sequestration in soil and biota (Follett et al., 2001). It is because of these numerous benefits that farmers are encouraged to participate in the CRP through incentive payments. Incentives include the government's sharing the cost of converting the cropland to alternate uses and making payments to land owners throughout the 10-year CRP contract.

Table 3.1  Global Land Use

| Land Use | 1985 | 1990 | 1995 | 1997 |
|----------|------|------|------|------|
| | Mha | | | |
| Total area | 13,399.8 | 13,399.7 | 13,387.0 | 13,387.0 |
| Land area | 13,043.7 | 13,043.6 | 13,048.4 | 13,048.4 |
| Arable land | 1,369.5 | 1,381.0 | 1,375.5 | 1,379.1 |
| Permanent crops | 105.4 | 119.0 | 129.6 | 131.3 |
| Others | 11,568.7 | 11,543.6 | 11,543.3 | 11,538.0 |

*Source*: Modified from FAO, *Production Yearbook*, vol. 52, FAO, Rome, Italy, 1998.

Table 3.2  Land Use in the U.S.

| Land Use | 1985 | 1990 | 1995 | 1997 |
|----------|------|------|------|------|
| | Mha | | | |
| Total area | 936.4 | 936.4 | 936.4 | 936.4 |
| Land area | 915.9 | 915.9 | 915.9 | 915.9 |
| Arable land | 187.8 | 185.7 | 176.9 | 176.9 |
| Permanent crops | 2.0 | 2.0 | 2.1 | 2.1 |
| Others | 726.1 | 728.1 | 736.9 | 736.9 |

*Source*: Modified from FAO, *Production Yearbook*, vol. 52, FAO, Rome, Italy, 1998.

## 3.2.5  Other Lands

This land use class encompasses all other land uses including wetlands, water bodies, urban lands, etc.

## 3.3 DATABASE

There are two types of databases. The national database on land use/land cover is kept by the United States Department of Agriculture (USDA). For the private sector of the U.S., this database is maintained by the National Resource Inventory (NRI) of the USDA. There are also global databases. Important among these is that of the Food and Agriculture Organization (FAO) of the United Nations. The data on global land use are published annually by FAO in *Production Yearbook* (FAO, 1998). Two examples of FAO's database on global and U.S. land use are shown in Table 3.1 and Table 3.2, respectively. The FAO database distinguishes only among arable land (cropland), permanent crops (perennials), and other lands.

## 3.4 LAND USE IN THE U.S.

The contiguous U.S. is divided into ten geographical regions (Figure 3.1). Each region has both federal land and nonfederal (private) land. Table 3.3 shows the amount of federal and nonfederal land in each of the ten regions in the U.S. There are six specific land use classes recognized in the nonfederal land. These are described in Section 3.2 and include cropland, pastureland, rangeland, forestland, CRP land, and

Table 3.3  Federal and Nonfederal Land in the Contiguous U.S.

| Region | Land Area | | Total |
|---|---|---|---|
| | Nonfederal | Federal | |
| | Mha | | |
| Northeast | 44.1 | 1.2 | 45.3 |
| Appalachia | 47.0 | 3.2 | 50.2 |
| Southeast | 47.0 | 2.8 | 49.8 |
| Lake states | 46.2 | 3.2 | 46.4 |
| Corn Belt | 65.2 | 1.6 | 65.8 |
| Delta states | 34.8 | 2.4 | 37.2 |
| Northern Plains | 76.1 | 2.4 | 78.5 |
| Southern Plains | 84.2 | 1.6 | 85.8 |
| Mountain states | 113.4 | 106.0 | 219.4 |
| Pacific states | 46.2 | 36.4 | 82.6 |
| Total | 604.2 | 160.8 | 765.0 |

Source: Modified from USDA, *The Second RCA Appraisal. Soil, Water, and Related Resources on Non-federal Land in the United States: Analysis of Conditions and Trends*, U.S. Government Printing Office, Washington, D.C., 1989, 280 pp.

**Figure 3.1**  Geographical regions of the conterminous U.S. (From *The Second RCA Appraisal. Soil, Water, and Related Resources on Non-Federal Land in the United States: Analysis of Conditions and Trends*, USDA, U.S. Government Printing Office, Washington, D.C., 1989).

other lands. The areas under five principal land uses (except CRP) for ten regions of the conterminous U.S. are shown in Table 3.4. It is apparent that large areas of cropland are located in the Corn Belt, Northern and Southern Plains, and Lake States regions. In contrast, rangelands are located in Southern Plains, and Mountain and Pacific regions. Forestlands are located in the Northeast, Appalachian, Southeast, Mountain

**Table 3.4   Land Use on Nonfederal Land in the U.S. in 1982**

| Region | Cropland | Pastureland | Nonfederal land Rangeland | Forestland | Other |
|---|---|---|---|---|---|
| | | | Mha | | |
| Northeast | 6.9 | 3.6 | 0 | 27.1 | 6.5 |
| Appalachian | 9.3 | 7.3 | 0 | 25.1 | 4.9 |
| Southeast | 7.3 | 4.9 | 1.6 | 26.7 | 6.5 |
| Lake states | 17.8 | 4.0 | <0.4 | 17.4 | 6.9 |
| Corn Belt | 37.2 | 10.1 | <0.4 | 10.5 | 6.9 |
| Delta states | 8.9 | 4.9 | <0.4 | 17.0 | 3.6 |
| Northern Plains | 37.7 | 3.2 | <1.6 | 0.8 | 4.5 |
| Southern Plains | 18.2 | 9.7 | 44.5 | 6.5 | 4.9 |
| Mountain | 17.4 | 2.8 | 74.5 | 10.9 | 7.3 |
| Pacific | 9.3 | 2.0 | 13.4 | 16.2 | 5.7 |
| Continental U.S. | 170.0 | 52.5 | 164.4 | 158.2 | 57.7 |

*Source*: Modified from USDA, *The Second RCA Appraisal. Soil, Water, and Related Resources on Non-federal Land in the United States: Analysis of Conditions and Trends*, U.S. Government Printing Office, Washington, D.C., 1989, 280 pp.

**Table 3.5   Land Use/Cover**

| Land Use | 1982 | 1987 | 1992 |
|---|---|---|---|
| | | Mha | |
| Cropland | | | |
| Cultivated (nonirrigated) | 128.8 | 122.9 | 112.5 |
| Cultivated (irrigated) | 19.5 | 19.2 | 19.3 |
| Noncultivated (nonirrigated) | 16.7 | 17.0 | 17.1 |
| Noncultivated (irrigated) | 5.5 | 5.6 | 5.9 |
| Subtotal | 170.5 | 164.7 | 154.8 |
| Pastureland | | | |
| Nongrazed | 13.2 | 8.9 | 5.2 |
| Grazed | 40.2 | 42.7 | 45.8 |
| Subtotal | 53.4 | 51.6 | 51.0 |
| Rangeland | | | |
| Nongrazed | 10.5 | 7.6 | 5.2 |
| Grazed | 155.0 | 155.4 | 156.4 |
| Subtotal | 165.5 | 163.0 | 161.6 |
| Forestland | | | |
| Nongrazed | 133.1 | 134.5 | 134.4 |
| Grazed | 26.6 | 25.7 | 25.5 |
| Subtotal | 159.7 | 160.2 | 159.9 |
| Minor land | 21.4 | 21.6 | 22.1 |
| Urban/built-up land | 21.0 | 23.5 | 26.5 |
| Rural transportation | 10.7 | 10.8 | 10.9 |
| Small water | 4.1 | 4.2 | 4.3 |
| Census water | 15.3 | 15.8 | 15.5 |
| Federal land | 163.9 | 164.3 | 165.2 |
| CRP | — | 5.6 | 13.8 |
| Total | 785.4 | 785.4 | 785.4 |

*Source*: From NRI, National Resource Inventory, Washington, D.C., 1982, 1987, 1992.

Table 3.6  Land Use in the U.S. over the 15-Year Period from 1982 to 1997

| Land Use | NRI | | | |
|---|---|---|---|---|
| | 1982 | 1987 | 1992 | 1997 |
| | Mha | | | |
| Cropland | | | | |
| Cultivated | 152.4 | 146.9 | 135.7 | 132.3 |
| Noncultivated | 18.0 | 17.7 | 19.1 | 20.3 |
| Total | 170.4 | 164.6 | 154.8 | 152.6 |
| Grazing land | | | | |
| Pastureland | 53.4 | 51.9 | 51.0 | 48.6 |
| Rangeland | 168.7 | 166.4 | 164.9 | 164.4 |
| Total | 222.1 | 218.3 | 215.9 | 213.0 |
| Forestland | 163.3 | 164.1 | 164.1 | 164.8 |
| Minor land | 20.1 | 20.2 | 20.5 | 20.7 |
| CRP land | — | 5.6 | 13.8 | 13.2 |
| Miscellaneous land | 209.5 | 212.6 | 216.3 | 221.1 |
| Total | 785.4 | 785.4 | 785.4 | 785.4 |

Source: From NRI, National Resource Inventory, Washington, D.C., 1997.

and Pacific study regions. The data in Table 3.4 are specifically chosen for 1982 because it does not show CRP land because the program was initiated by the 1985 Farm Bill.

Land use change in the U.S. between 1982 and 1992 can be deduced from the data in Table 3.5 (NRI 1981; 1987; 1992). Total cropland area decreased from 170.5 Mha in 1982 to 154.8 Mha in 1992, pastureland area decreased from 53.4 Mha in 1982 to 51.0 Mha in 1992, and rangeland area decreased from 165.5 Mha in 1982 to 161.6 Mha in 1992. There was no change in the area under forestland. There were two land uses that registered drastic increases. The area under urban/built-up land was 21.4 Mha in 1982, 23.5 Mha in 1987, and 26.5 Mha in 1992. There was also a drastic increase in the cropland area converted to CRP. The area under CRP was only 5.6 Mha in 1987 compared with 13.8 Mha in 1992 (Table 3.5).

The data in Table 3.6 is a continuation and simplification of the information contained in Table 3.5. There are fewer land use classes in Table 3.6 than in Table 3.5. The trends in prominent land use classes were the same in 1997 as that in 1992. There was a continuous decline in land area devoted to cropland, pastureland, and rangeland. Decline in these land uses was due to conversion to other land uses, with about 60% of the land going into CRP and 40% into miscellaneous lands, but mainly urban.

## 3.5 TRENDS IN CROPLAND

Trends in cropland area are shown in Figure 3.2, Figure 3.3, Figure 3.4, and Figure 3.5. The data show that the 20th century can be divided into six distinct periods. Period 1 is from 1910 to about 1920, during which the area under cropland increased from 133 Mha to 150 Mha. Period 2 is from 1920 to about 1955, during

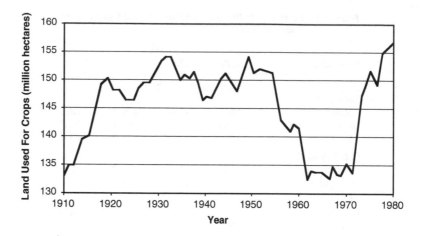

**Figure 3.2**   Cropland area in actual production. (Redrawn from Batie, S., in *Using our Natural Resources, 1983 Yearbook of Agriculture*, USDA, Washington, D.C. 1983, 404–413.)

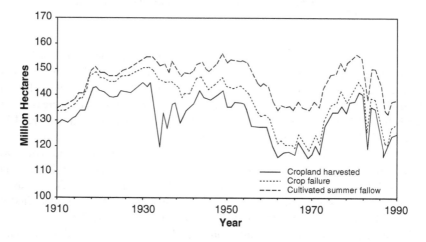

**Figure 3.3**   Trends in cropland area, 1910 to 1990. While using cropland harvested as the base area, other categories are incrementally added. (Redrawn from Heimlich and Daugherty, in *Agriculture and the Environment, The 1991 Yearbook of Agriculture*, USDA, Washington, D.C., 1991, 3-9.)

which the area under cropland remained relatively stable at about 150 Mha (between 147 Mha and 155 Mha). Period 3 is from 1952 to 1962 when the area under cropland decreased to about 133 Mha, the same as prior to 1910. Period 4 is from 1962 to 1972 when the cropland area remained stable at about 133 Mha. Period 5 is from 1972 to the 1980s when the area increased again to about 155 Mha. Combining these five periods into one shows overall trends in cropland between 1910 and 1980. The data in Figure 3.3 indicate the amount of cropland under different categories between 1910 and 1990 (Heimlich and Daugherty, 1991). For example, the cropland area of 133.6 Mha in 1910 was expanded to 151.8 Mha during World War I and

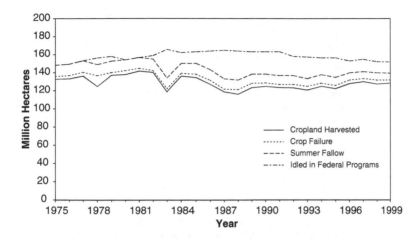

**Figure 3.4**   Trends in cropland area from 1975 to 1999. While using cropland harvested as
the base area, other categories are incrementally added. (Redrawn from Heimlich
and Daugherty, *Agriculture and the Environment, 1991 Yearbook of Agriculture*,
USDA, Washington, D.C., 1991, 3-9.)

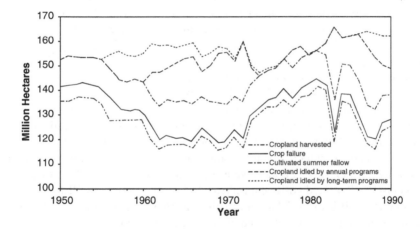

**Figure 3.5**   Trends in cropland area, 1950 to 1990. While using cropland harvested as the
base area, other categories are incrementally added. (Redrawn from Heimlich
and Daugherty, *Agriculture and the Environment, 1991 Yearbook of Agriculture*,
USDA, Washington, D.C., 1991, 3-9.)

maintained until after World War II. There was another expansion in the cropland
area during the 1970s to 153.8 Mha in response to export boom. However, the
cropland area decreased to 133.6 Mha because of the downturn in export during the
mid-1980s and conversion to CRP. Consequently, the cropland area in the mid-
1980s was similar to that in 1910. While the overall cropland area remained constant,
there were considerable regional changes in cropland area between 1949 and 1987
(Table 3.7). There was an increase in the cropland area by 8.1 Mha in the Corn Belt,
Northern Plains, and Rocky Mountain regions. In the remaining regions, there was

Table 3.7    Regional Changes in Cropland Area
between 1949 and 1987

| Region | Change in Cropland Area (Mha) |
|---|---|
| Northeast | −3.6 |
| Lake states | −1.6 |
| Corn Belt | +1.5 |
| Northern Plains | +3.6 |
| Appalachian | −3.2 |
| Southeast | −3.9 |
| Delta states | −0.2 |
| Southern Plains | −0.7 |
| Rocky Mountains | +3.0 |
| Pacific | −0.7 |
| U.S. | −5.8 |

*Source*: Recalculated from Heimlich and Daugherty, *Agriculture and the Environment, 1991 Yearbook of Agriculture*, USDA, Washington, D.C., 1991, p. 3–9.

a decrease in cropland area by 13.9 Mha. Thus, there was a net loss in the cropland area in the U.S. by 5.8 Mha between 1949 and 1987 (Table 3.7).

The sixth period is the last 25 years of the 20th century (Figure 3.4). In this time span, there was a consistent decline in cropland area during the last two decades, and over the 15-year period between 1982 and 1997, there was a decrease in the cropland area by 17.8 Mha or 10.4%, and a decrease in the grazing land area by 9.1 Mha or 4.1%. There was a slight increase in the forestland by 1.5 Mha but a drastic conversion from cropland into CRP of 13.2 Mha (Table 3.6).

A major contributor to the decline in area under cropland during the last 15 years of the 20th century was implementation of CRP as a result of the Farm Bills of 1985, 1990, and 1996. These bills discouraged producing an agricultural commodity on HEL without following an approved conservation plan. The disincentive was in the form of no price support, no farm storage facility loans, no federal crop insurance, no disaster payments, etc. Farmers were required to meet strict erosion control goals. In contrast, CRP offered positive incentives to farmers for retiring HEL for 10 years. For converting HEL to CRP, farmers received annual rental payments and assistance in establishing trees or any other permanent vegetation cover.

## 3.6 TRENDS IN GRAZING LAND

Management of grazing land is important to controlling sediment load, reducing wind erosion, improving water quality, decreasing emission of greenhouse gases, and protecting wildlife habitats. The land area under grazing land is estimated at 236 Mha including rangeland, pastureland, and grazed forestland (NRI, 2000). Together, pastureland and rangeland amount to nearly 212 Mha or about 35% of the nonfederal land. Total pastureland and rangeland has declined by 10.5 Mha since 1982. Between 1992 and 1997, the combined decline in pastureland and rangeland use was about 4.0 Mha. About 10.5 Mha was converted from pastureland and rangeland to other uses, while 6.5 Mha was converted to pastureland and rangeland.

Development (e.g., urbanization) accounted for nearly 18% of decreases in pasture-land and rangeland. Thus, most of the change in pastureland and rangeland was conversion to or from cropland. About 24% of the decrease in pastureland and rangeland resulted in increases in cultivated cropland. In comparison, 45% of the increase in pastureland and rangeland occurred due to conversion of cropland to pasture or rangeland (NRI, 2000).

Distribution of the area under different land uses for each state in the contermi-nous U.S. for 1992 is shown in Table 3.8 (NRI, 1992). As one would expect, cropland is primarily located in the Northern Plains, Corn Belt, and the Lake States regions. Rangeland is mostly concentrated in the Southern Plains and Mountain States regions. Forestland occurs in the Appalachian, Southeast, Pacific states, Northeast regions, Delta states, and Lake states (Table 3.4).

## 3.7 CONCLUSIONS

Land use trends in the U.S. during the 20th century reflect the impact of market forces and a heightened awareness of the environment. The cropland area increased during the World War eras and in the 1970s because of increased demand for crops for both domestic consumption and export. The decline in cropland area during the 1980s was due to downturn in exports of food products during the 1990s because of the concern of nonpoint source pollution caused by erosion from HELs. In addition to conversion to CRP, there has also been an increase in the area under forestland.

Future trends in land use, especially in the areas under agricultural and forestry land uses, will be governed by environmental issues and the demands of an increas-ing population. In addition to the concern about water quality and nonpoint source pollution, there is a growing awareness about the accelerated greenhouse effect and the potential of soil resources through C sequestration to mitigate it. Conversion of the HEL to CRP and increases in the area under forestlands can lead to C sequestration in the soils and biomass and decrease in the rate of enrichment of the atmospheric concentrations of several GHGs (e.g., $CO_2$, $CH_4$, $N_2O$, $NO_x$) (Lal et al., 1998; Follett et al., 2001; Kimble et al., 2002). The future Farm Bills may provide incentives to land owners for C sequestration in soil and biomass through conversion of cropland to less intensive land uses, and adoption of management practices to remove $CO_2$ from the atmosphere and reducing the risks of climate change.

An important factor determining the trends in land use during the 21st century is the quest for developing agricultural systems that are environmentally friendly. Growing concerns about pollution and contamination of natural waters and risks of the climate change will lead to creating legislation that influences land use and adoption of environment-friendly soil–crop–vegetation management practices. Rather than a cause, agriculture can be a solution to the environmental issues. It is this realization that will determine the future trends in land use during this century.

Table 3.8  Land Cover in the U.S. by States (1000 ha)

| State | Cropland | | Federal Land | Forestland | Misc. Land | Pastureland | Rangeland | Rural Transport | Water Census | Total |
|---|---|---|---|---|---|---|---|---|---|---|
|  | Cultivated | Noncultiv. |  |  |  |  |  |  |  |  |
| AL | 1134.6 | 138.9 | 372.9 | 8485.6 | 487.5 | 1521.4 | 27.3 | 827.9 | 395.5 | 13391.6 |
| AZ | 399.1 | 85.6 | 12253.9 | 1909.3 | 1127.2 | 30.9 | 13041.9 | 568.2 | 110.2 | 29526.2 |
| AR | 3004.5 | 123.7 | 1298.0 | 5773.8 | 242.0 | 2317.6 | 64.2 | 535.1 | 416.7 | 13775.5 |
| CA | 2578.2 | 1489.7 | 18936.4 | 5986.9 | 1920.5 | 469.8 | 6936.2 | 2023.7 | 763.7 | 41105.1 |
| CO | 3073.5 | 544.6 | 9681.4 | 1519.7 | 1238.1 | 508.1 | 9525.1 | 685.3 | 184.1 | 26959.7 |
| CT | 38.0 | 54.5 | 5.9 | 712.3 | 56.5 | 44.5 | 0.0 | 330.0 | 58.0 | 1299.7 |
| DE | 199.6 | 2.4 | 13.4 | 142.7 | 52.7 | 10.4 | 0.0 | 82.9 | 25.4 | 529.5 |
| FL | 683.4 | 529.6 | 1534.3 | 5009.3 | 1030.1 | 1769.7 | 1403.0 | 1879.9 | 1354.7 | 15194.0 |
| GA | 1904.6 | 188.8 | 844.4 | 8787.5 | 631.4 | 1244.5 | 0.0 | 1245.2 | 411.3 | 15257.7 |
| ID | 1764.8 | 501.6 | 13475.4 | 1628.4 | 566.2 | 503.1 | 2698.3 | 237.7 | 267.9 | 21643.3 |
| IL | 9380.7 | 372.3 | 210.6 | 1383.6 | 562.5 | 1118.4 | 0.0 | 1252.2 | 313.2 | 14593.4 |
| IN | 5162.4 | 306.0 | 197.2 | 1467.5 | 480.6 | 755.3 | 0.0 | 847.7 | 155.4 | 9372.1 |
| IA | 9456.0 | 656.3 | 74.5 | 781.6 | 1195.1 | 1502.1 | 0.0 | 719.9 | 189.9 | 14575.4 |
| KS | 10079.4 | 671.3 | 245.1 | 538.8 | 1442.9 | 933.3 | 6362.8 | 808.3 | 228.0 | 21310.0 |
| KY | 1374.2 | 686.4 | 486.0 | 4173.2 | 432.6 | 2371.0 | 0.0 | 668.9 | 273.8 | 10466.1 |
| LA | 2339.7 | 77.0 | 511.4 | 5245.3 | 1287.3 | 918.1 | 91.7 | 713.9 | 1183.4 | 12367.7 |
| ME | 64.6 | 116.5 | 66.2 | 7104.9 | 285.3 | 45.0 | 0.0 | 282.1 | 651.0 | 8615.7 |
| MD | 587.9 | 89.2 | 67.4 | 956.7 | 144.1 | 220.7 | 0.0 | 443.2 | 200.0 | 2709.2 |
| MA | 28.9 | 81.3 | 36.1 | 1124.3 | 125.5 | 68.7 | 0.0 | 529.8 | 151.0 | 2145.6 |
| MI | 2756.9 | 879.3 | 1281.0 | 6316.3 | 974.4 | 952.4 | 0.0 | 1491.6 | 506.6 | 15158.6 |
| MN | 7586.5 | 1055.9 | 1369.0 | 5590.8 | 2517.9 | 1328.2 | 0.0 | 978.4 | 1433.5 | 21860.2 |
| MS | 2200.9 | 116.4 | 698.5 | 6379.9 | 452.1 | 1637.6 | 0.0 | 541.2 | 325.2 | 12351.6 |
| MO | 4337.3 | 1064.3 | 816.3 | 4717.2 | 943.9 | 4820.3 | 51.2 | 945.5 | 355.8 | 18051.7 |
| MT | 4971.4 | 1113.0 | 10975.9 | 2086.5 | 1685.5 | 1363.9 | 14906.7 | 443.4 | 538.7 | 38085.1 |
| NE | 7055.1 | 730.8 | 298.9 | 314.4 | 863.9 | 835.9 | 9173.8 | 506.5 | 255.8 | 20035.1 |
| NV | 68.7 | 239.8 | 24398.6 | 142.7 | 147.3 | 120.3 | 3178.5 | 159.6 | 179.9 | 28635.4 |
| NH | 8.4 | 48.9 | 302.5 | 1591.3 | 87.8 | 39.8 | 0.0 | 227.9 | 96.6 | 2403.2 |
| NJ | 190.9 | 72.0 | 64.5 | 714.7 | 156.2 | 64.4 | 0.0 | 642.6 | 111.6 | 2016.9 |
| NM | 577.8 | 187.7 | 11086.2 | 1861.7 | 1147.2 | 85.7 | 16103.3 | 350.5 | 92.6 | 31492.7 |

| | | | | | | | | | | |
|---|---|---|---|---|---|---|---|---|---|---|
| NY | 919.9 | 1352.9 | 93.4 | 6951.9 | 399.6 | 1214.6 | 0.0 | 1216.1 | 570.8 | 12719.0 |
| NC | 2241.2 | 170.6 | 990.6 | 6466.7 | 394.5 | 817.2 | 0.0 | 1433.6 | 1127.1 | 13641.4 |
| ND | 9106.6 | 906.7 | 789.5 | 172.3 | 1691.9 | 472.6 | 4178.5 | 544.0 | 449.9 | 18312.0 |
| OH | 4116.3 | 711.1 | 151.9 | 2680.8 | 515.9 | 918.0 | 0.0 | 1439.8 | 170.4 | 10704.3 |
| OK | 3890.5 | 189.0 | 486.4 | 2827.9 | 676.4 | 3124.1 | 5690.2 | 758.8 | 475.3 | 18118.7 |
| OR | 1079.8 | 448.1 | 13067.8 | 4790.9 | 462.0 | 768.7 | 3793.8 | 455.4 | 275.5 | 25142.0 |
| PA | 1183.6 | 1081.0 | 276.0 | 6198.2 | 462.4 | 941.4 | 0.0 | 1388.9 | 203.4 | 11734.9 |
| RI | 2.5 | 7.6 | 1.7 | 159.0 | 12.0 | 9.8 | 0.0 | 76.8 | 44.7 | 314.0 |
| SC | 1114.6 | 92.4 | 467.9 | 4420.1 | 408.7 | 481.5 | 0.0 | 751.1 | 321.8 | 8058.2 |
| SD | 5626.6 | 1025.0 | 1176.4 | 218.6 | 1318.0 | 873.3 | 8876.0 | 459.4 | 399.8 | 19973.1 |
| TN | 1442.8 | 522.7 | 558.1 | 4686.3 | 395.5 | 2090.2 | 0.0 | 874.5 | 345.2 | 10915.3 |
| TX | 11122.5 | 314.6 | 1296.1 | 4030.9 | 2577.6 | 6762.3 | 38103.7 | 3331.0 | 1564.8 | 69103.4 |
| UT | 327.0 | 407.5 | 14399.5 | 658.1 | 869.4 | 269.2 | 4067.3 | 227.2 | 763.9 | 21989.0 |
| VT | 47.9 | 208.9 | 149.1 | 1674.7 | 30.3 | 141.3 | 0.0 | 131.2 | 106.6 | 2490.0 |
| VA | 702.7 | 471.3 | 966.8 | 5479.1 | 286.2 | 1393.8 | 0.0 | 883.4 | 375.3 | 10558.6 |
| WA | 2266.9 | 462.8 | 5050.1 | 5077.8 | 793.2 | 547.2 | 2216.2 | 748.9 | 484.9 | 17647.9 |
| WV | 80.5 | 289.7 | 485.9 | 4263.1 | 158.2 | 651.1 | 0.0 | 279.0 | 68.6 | 6276.0 |
| WI | 2852.1 | 1523.9 | 740.2 | 5426.9 | 1300.0 | 1195.6 | 0.0 | 953.7 | 551.3 | 14543.7 |
| WY | 372.7 | 546.6 | 394.7 | 394.7 | 529.7 | 364.4 | 10527.9 | 218.8 | 229.2 | 25332.8 |
| Total | 131504.3 | 22956.0 | 164897.9 | 159024.7 | 35565.9 | 50637.4 | 161017.4 | 37140.6 | 19757.9 | 782502.1 |

*Source:* From NRI, National Resource Inventory, Washington, D.C., 1992a.

# REFERENCES

Batie, S. How farmers affect natural resources. In *Using Our Natural Resources, 1983 Yearbook of Agriculture*. USDA, Washington, D.C., 1983, pp. 404–413.

Dukes, D. CRP: A wakeup call for agriculture. *J. Soil Water Cons.* 51, 140–141, 1996.

FAO. Forest Resources Assessment 1990, Guidelines for Assessment. FAO, Rome, 1990.

FAO. *Production Yearbook*, Vol. 52. FAO, Rome, 1998.

Follett, F.R., E.G. Pruessner, S.E. Samson-Liebig, J.M. Kimble, and S.W. Waltman. Carbon sequestration under the Conservation Reserve Program in the historic grassland soils of the United States of America. In *Soil Carbon Sequestration and the Greenhouse Effect*. R. Lal (Ed.), Soil Sci. Soc. Amr. Special Publ. #57, Madison, WI, 2001, pp. 27–40.

Graetz, D. and Grasslands. In *Changes in Land Use and Land Cover: A Global Perspective*. W.B. Meyer and B.L. Turner (Eds.), Cambridge University Press, Cambridge, 1994, pp. 125–147.

Heimlich, R.E. and A.B. Daugherty. America's cropland: where does it come from? In *Agriculture and the Environment, 1991 Yearbook of Agriculture*. USDA, Washington, D.C., 1991, pp. 3–9.

Kimble, J.M., R. Lal, and R.F. Follett (Eds.). *Agricultural Practices and Policies for Carbon Sequestration in Soil*. CRC/Lewis Publishers, Boca Raton, FL, 2002, 512 pp.

Lal, R., J.M. Kimble, R.F. Follett, and C.V. Cole. *The Potential of U.S. Cropland to Sequester Carbon and Mitigate the Greenhouse Effect*. Ann Arbor Press, Chelsea, MI, 1998, 128 pp.

Lillesand, T.M. and R.W Kiefer. *Remote Sensing and Image Interpretation*. J. Wiley & Sons, New York, 1987.

NRI. National Resource Inventory, Washington, D.C., 1982.

NRI. National Resource Inventory, Washington, D.C., 1987.

NRI. National Resource Inventory, Washington, D.C., 1992a.

NRI. Instructions for collecting 1992 National Resources Inventory Sample Data. USDA-NRCS, Washington, D.C., 1992b.

NRI. *National Range and Pasture Handbook*. Grazing Land Technology Institute. USDA-NRCS, Washington, D.C., 1997.

NRI. National Resources Inventory: Background and Highlights. USDA-NRCS, Washington, D.C., 2000.

Sobecki, T.M., D.L. Moffitt, J. Stone, C.D. Franks, and A.G. Mendenhall. A broad scale perspective on the extent, distribution and characteristics of U.S. grazing land. In *The Potential of U.S. Grazing Lands to Sequester Carbon and Mitigate the Greenhouse Effect*. R.F. Follett, J.M. Kimble, and R. Lal (Eds.), CRC Press, Boca Raton, FL, 2001, pp. 21–63.

Turner, B.L. and W.B. Meyer. Global land use and land cover change: an overview. In *Changes in Land Use and Land Cover: A Global Perspective*. W.B. Meyer and B.L. Turner (Eds.), Cambridge University Press, Cambridge, 1994, pp. 3–10.

Uri, N.D. Agriculture and the environment — the problem of soil erosion. *J. Sust. Agric.* 16, 71–94, 2000.

USDA. *The Second RCA Appraisal. Soil, Water, and Related Resources on Non-federal Land in the United States: Analysis of Conditions and Trends*. USDA, U.S. Govt. Printing Office, Washington, D.C., 1989, 280 pp.

Williams, M. Forests and tree cover. In *Changes in Land Use and Land Cover: A Global Perspective*. W.B. Meyer and B.L. Turner (Eds.), Cambridge University Press, Cambridge, 1994, pp. 97–124.

CHAPTER **4**

# Soil Degradation Category Indicators Methodology

## 4.1 SOIL DEGRADATION

A fundamental approach toward assessing land quality is to document the area of land of a given land cover or use type (grazing land, cropland, forestland, etc.) affected by degradative processes that are outside of the normal range of variation or are out of balance with ameliorative process. This geospatial approach is consistent with concepts in the Santiago Agreement, which established criteria and indicators for sustainability of the world's forest resources (Coulombe, 1995). Many of the criteria and indicators developed for use under the Santiago Agreement are area based (i.e., absolute area or the percentage of forestland affected by process or agents beyond the range of historic variation, such as excessive erosion) (Coulombe, 1995). A fundamental geospatial approach to land degradation assessment is also proposed by Dumanski (1997) in establishing a series of five "land quality indicators," three of which specifically incorporate the areal extent or pattern of land conditions: land use intensity, land use diversity over the landscape, and extent, duration, and timing of vegetative cover on the land.

Following the above concepts, we established criteria to enable computation of simple indicators of land degradation processes from the National Resources Inventory (NRI) data (NRCS, 1994; NRCS, 1999). Acreage of U.S. land area affected by various magnitudes or degrees of several degradative processes, as indicated by the presence and relative degree of expression of respective indicators, was tabulated. Categories of the relative degree or intensity of land degradation of light, moderate, strong, and extreme were adapted from FAO (1994). These categories are used to rank the expression of a specific degradation process relative to a nondegraded, relatively productive state of the land suitable, within natural limitations, for agricultural or other extensive land use.

45

The acreage tabulations are presented by the NRI broad land cover and use classes of cropland-cultivated, cropland-noncultivated, pastureland, rangeland, forestland, and miscellaneous land cover and uses, respectively (SCS, 1992; NRCS, 1997). Differences in data collection methods and procedures for rangeland vs. other broad land cover/uses and for some 1997 NRI data vs. that of previous years (1992 and earlier) necessitated modification of criteria as a function of broad land cover and use category and year of data collection. Accordingly, rangeland data are reported up to and including 1992, but little beyond that extent are available for 1997. Criteria for vegetative condition/status of rangeland were set to follow as closely as possible those used in the State of the Land Report covering the nations nonfederal rangeland (Spaeth, et al. 1998).

### 4.1.1   Land Degradation Severity Classes — Definitions

*Light* — Land/terrain is suitable for agricultural use (including grazing), requiring only modification of management to maintain or restore full productivity; original biotic function still largely intact.

*Moderate* — Land/terrain is still suitable for agricultural use (including grazing) but productivity significantly impacted; requires land treatment over and above mere modification of management systems to restore full productivity and restore original biotic function.

*Severe* — Land/terrain, in terms of respective degradation process, is largely beyond restoration at a farm or ranch level through simple management or other on-farm land treatment measures; requires more extensive treatment that may include major structural and engineering modification of the land or area-wide mitigation or restoration measures to restore productivity and biotic function.

*Extreme* — Land/terrain, in terms of the respective degradation process, is beyond restoration in the sense that drastic measures are required to reestablish landscape components (soil, aquatic systems, vegetation, etc.) needed to have some level of agricultural productivity and biotic function.

### 4.1.2   Class Criteria

The following are the specific criteria used to query the NRI database and place acreage of U.S. land into the four land degradation severity classes defined above. Separate placement was made for each of the following degradative processes: sheet and rill water erosion, wind erosion, soil salinization, mining or other geologic resource extraction, concentrated flow water erosion on rangeland, and vegetative decline on rangelands. The capital letter following each degradative process indicates the general land degradation process to which the specific process belongs: E = erosion, S = salinization, M = mining or related resource extraction activities, V = decline in vegetative cover or quality (i.e., ecological condition).

## 4.2 WATER EROSION, SHEET AND RILL (E) — DEGRADATION OF LAND THROUGH THE PROCESS OF SHEET AND RILL SOIL EROSION BY WATER

*Light* — average annual soil loss, as computed from the Universal Soil Loss Equation (USLE) (Wischmeier and Smith, 1978) and expressed as T/ha/yr, is less than the soil T factor.*

*Moderate* — average annual soil loss <2T

*Severe* — average annual soil loss between 2T and 11T

*Extreme* — average annual soil loss >11T

## 4.3 WATER EROSION, CONCENTRATED FLOW (E) — DEGRADATION OF LAND THROUGH THE PROCESS OF CONCENTRATED FLOW EROSION (I.E., GULLYING, STREAMBANK EROSION, ETC.) (APPLIED ONLY TO THE BROAD LAND COVER AND USE CATEGORY OF RANGELAND)

*Light* — gully or streambank or concentrated flow erosion "stable" and soil T factor of 11 (i.e., no active erosion on soil loss-tolerant soils, some relict erosion but concentrated flow erosional process stabilized).

*Moderate* — gully or streambank or concentrated flow erosion "slight" and soil T factor of 5 (i.e., some active erosion on soil loss-tolerant soils)

*Severe* — (gully or steambank or concentrated flow erosion "significant" and soil T factor <11) or (gully or steam bank or concentrated flow erosion "slight" and soil T factor <11) (i.e., major active erosion on soil loss-tolerant soils or some active erosion on soils intolerant of soil loss).

*Extreme* — gully or streambank or concentrated flow erosion "significant" and soil T factor <11 (i.e., major active erosion on soils intolerant of soil loss).

## 4.4 WIND EROSION (E) — DEGRADATION OF LAND THROUGH SOIL EROSION BY WIND

Same criteria as for sheet and rill erosion by water, but soil loss was computed from the wind erosion equation (WEQ).

---

* The soil loss tolerance factor (soil T factor) is a soil-specific sustainable level of erosion traditionally expressed in units of tons per acre per year. At rates of soil erosion below this value, soil productivity and quality can be maintained. In U.S. soils surveys (Soil Survey Division Staff, 1993), soil T factors are expressed in English units of tons per acre per year. Most deep, productive agricultural soils have a soil T factor of 5 in English units. The corresponding value in metric units of megagrams per hectare per year is 11. Criteria initially established for this class had included a requirement for a conservation treatment need of "erosion control" indicated in the NRI database as a secondary indirect indicator of excessive soil erosion. However, lack of treatment needs data in the 1997 NRI prevented trending between 1992 and 1997. Therefore, the treatment needs criteria were dropped for all years. Conservation treatment needs in the 1992 and earlier NRI database are "based on the judgment of a qualified specialist as guided by the local technical guide, prevailing operations, and the guides used in the development of conservation plans" (NRCS, 1992).

## 4.5 SALINIZATION (S) — DEGRADATION OF LAND THROUGH SALT ACCUMULATION*

For 1992 and earlier NRI data:

*Light* — maximum soil salinity, expressed as electrical conductivity of a saturated paste extract, >0 and ≤2 dS/m and broad land cover/use of cropland or noncultivated cropland and conservation treatment needs of drainage or irrigation management, or the land is irrigated.*

*Moderate* — maximum soil salinity >0 and ≤2, and conservation treatment needs of toxic salt reduction.

*Strong* — maximum soil salinity >2 and ≤4 and conservation treatment needs of toxic salt reduction.

*Extreme* — maximum soil salinity >4 and conservation treatment needs of toxic salt reduction.

For 1997 NRI data:

*Light* — maximum soil salinity >0 and ≤2, and broad land cover/use of either cropland or noncultivated cropland and land is irrigated or conservation practices irrigation system (447) or surface drainage (607 or 608) (NHCP, 2002) is applied.

*Moderate* — maximum soil salinity >0 and ≤2, and saline deposits are visibly present.

*Strong* — maximum soil salinity >2 and ≤4, and saline deposits are visibly present.

*Extreme* — maximum soil salinity >4 and saline deposits are visibly present.

---

* Changes in 1997 data collection protocols, namely, the lack of conservation treatment needs data and the substitution of salt deposit data, necessitated the development of two sets of query logic to adequately categorize soil salinization. As a result, data for the 1997 year may not be comparable in terms of observed trends in acreage with that of previous NRI years (i.e., 1992 and earlier). The overall query logic for both data sets (i.e., 1992 and 1997) identified a light category based on potential for soil salinization due to management and land treatment on cropland that is currently not of excessive soil salinity, thus amenable to treatment by simple on-farm management modification. Soil salinity levels in the NRI database alone are not used as sole criteria, as soil salinity levels reported in the NRI are derived from soil survey data that reflect native or natural salt levels as opposed to levels that have been anthropogenically accentuated (Soil Survey Division Staff, 1993).

The light category in the 1992 and earlier NRI data was defined using treatment needs data of drainage or irrigation water management or the presence of irrigation on low but nonzero salinity cropland. This represents a set of conditions indicating the potential for increased or excessive salinization. Such potential is not readily derivable for the 1997 NRI data, therefore the light category for the 1997 NRI was defined using the practice application data recorded that year as a substitute for the treatment needs data used in previous NRI years. It is still centered on the concept of soils with some nonzero but low native salt content that are being cropped and thus have the potential for salt accumulation due to poor drainage or the presence of irrigation.

Irrigation is not a requirement to be placed in the light category, as some soils with high water tables, even in humid regions, experience elevated salinity levels after vegetation removal or application of cropping practices that significantly change the hydrology. Irrigated cropland with actual evidence of salt accumulation (1997 data) or a specified treatment need of salt reduction (1992 and earlier) is relegated to the MODERATE category in both sets of criteria, as the intent of this class is to include land where actual reclamation measures may be required to restore full productivity. Irrigation water management treatment needs was included in the 1992 criteria to be consistent with implications in 1997 NRI instructions (NRCS, 1997) that imply areas with salt deposits would need conservation treatments of toxic salt reduction, irrigation water management, and soil salinity management.

## 4.6 MINING OR OTHER SOIL REMOVAL/DESTRUCTION (M) — DEGRADATION OF LAND THROUGH SOIL DESTRUCTION

*Light* — category not applicable

*Moderate* — broad land cover/use of cropland cultivated or noncultivated, pastureland, rangeland, or forestland *and* applied conservation practices of land reconstruction, abandoned (543) or land reconstruction, current (544) (NHCP, 2002) *and* a primary land use of business/commercial mineral extraction (NRCS, 1992; 1997).

*Severe* — broad land cover/use of cropland cultivated or noncultivated, pastureland, rangeland, or forestland and a primary land use of business/commercial mineral extraction.

*Extreme* — broad land cover/use of "barren land" and land use of strip mines, quarries, gravel pits, etc. (NRI code 613) or oil wasteland (NRI code 619) (NRCS, 1992).

## 4.7 VEGETATIVE CONDITION/STATUS (V) — DEGRADATION OF LAND THROUGH DEGRADATION OF THE QUANTITY, QUALITY, OR ECOLOGICAL STATUS OF VEGETATIVE LAND COVER (APPLIED ONLY TO THE BROAD LAND COVER/USE CATEGORY OF RANGELAND)

*Light* — range condition "good" and range trend "stable" or "down," or range condition "fair" and range trend "up" or only conservation treatment need is management for forage improvement with no noxious weeds present.*

*Moderate* — range condition "fair" and range trend "stable" or "down," or range condition "poor" and range trend "up" or conservation treatment need is management for forage improvement and other conservation treatment needed of drainage or toxic salt reduction to maintain optimal plant growth conditions.

*Severe* — range condition "poor" and range trend "stable" or conservation treatment needed of weed control/brush management or noxious weeds are present.*

*Extreme* — range condition "poor" and range trend "down" or conservation treatment needs of either plant/forage reestablishment or brush management.

## REFERENCES

Coulombe, M.J. Sustaining the world's forests: the Santiago agreement. *J. Forestry* 4, 18–21, 1995.

Dumanski, J. Criteria and indicators of land quality and sustainable land management. *IIC Journal* 3/4, 1997.

* Rangeland ecological status (Joyce, 1989) has traditionally been indicated in the NRI by application of the range condition concept (NRCS, 1992). Presence of noxious weeds was used as criteria for placement in the "severe" land degradation category, as control often requires area or state-wide programs for weed control or eradication (i.e., beyond simple individual farm or ranch-level management).

FAO. Land degradation in south Asia: its severity, causes and effects upon the people. World
        Soil Resources Report no. 78. United Nations Food and Agriculture Organization,
        Rome, 1994.

Joyce, L.A. An analysis of the range forage situation in the United States: 1989–2040. A
        technical document supporting the 1989 USDA Forest Service RPA assessment.
        USDA Forest Service General Technical Report RM-180. Rocky Mountain Forest
        and Range Experiment Station, Fort Collins, CO, 1989.

NHCP. *National Handbook of Conservation Practices.* GM 450-VI. United States Department
        of Agriculture, Washington, D.C., 2002. (http://www.ftw.nrcs.usda.gov/nhcp_2.html)

NRCS. Instructions for Collecting 1992 National Resources Inventory Sample Data. United
        States Department of Agriculture—Natural Resources Conservation Service. Washing-
        ton, D.C., 1992.

NRCS. Summary Report 1992 National Resources Inventory. United States Department of
        Agriculture—Natural Resources Conservation Service. Washington, D.C., 1994.

NRCS. Summary Report 1997 National Resources Inventory. United States Department of
        Agriculture—Natural Resources Conservation Service. Washington, D.C., 1999.

NRCS. Instructions for Collecting 1992 National Resources Inventory Sample Data. United
        States Department of Agriculture—Natural Resources Conservation Service. Washing-
        ton, D.C., 1997.

SCS. Instructions for Collecting 1992 National Resources Inventory Sample Data.
        USDA—Natural Resources Conservation Service (formerly Soil Conservation Ser-
        vice), Washington, D.C., 1992.

Soil Survey Division Staff. *Soil Survey Manual. Agriculture Handbook No. 18.* United States
        Department of Agriculture, Washington, D.C., 1993.

Spaeth, K.E., F.B. Pierson, L.J. Rich, F. Busby, P.L. Shaver, and A.G. Mendenhall. State of
        the Nation's nonFederal Rangeland: 1992 Natural Resource Inventory Summary —
        Nation, Regions, and States. NRCS Technical Note, 1998.

Wischmeier, W.H., and D.D. Smith. *Predicting rainfall erosion losses — a guide to conservation
        planning. Agric. Handbook No. 537.* U.S. Department of Agriculture, Washington, D.C.,
        1978.

# SECTION II

# Water and Wind Erosion

CHAPTER 5

# Soil Erosion by Water from Croplands

Accelerated erosion driven by anthropogenic forces is the most widespread soil degradation process in the U.S. It is a physical process that results in permanent loss of effective rooting depth, and pollution of waters and emission of trace gases or greenhouse gases (GHGs) from soil. Whereas natural erosion is a constructive process that leads to formation of fertile soils around the world (river deltas, for example) and provides nutrients to water bodies for the growth of aquatic biota, accelerated erosion is a severe degradative process with adverse impact on productivity and the environment. Accelerated erosion is exacerbated by anthropogenic activities involving conversion of natural ecosystems to farmland. The extent, rate, and total magnitude of erosion are accentuated by deforestation, biomass burning, plowing and discing, cultivation of open-row crops, etc. Soil in unprotected tilled fields without any vegetal cover (Plate 5.1) may lead to the development of a surface seal and crust, high soil erodibility due to weak soil structure and low water infiltration rate (Plate 5.2), and concentrated flow (Plate 5.3) resulting in severe soil erosion. Among the principal land use classes (Chapter 3), cropland is the most susceptible to erosion by water because many farming practices remove the protective vegetative cover.

## 5.1 CROPLAND AREA AFFECTED BY WATER EROSION

The cropland area affected by water erosion on private U.S. land is shown in Table 5.1 and Figure 5.1 using the data from the National Resource Inventory (NRI) from 1982 to 1997. As reported in Chapter 3, the area in cultivated cropland is progressively decreased over this period from 152.4 Mha in 1982 to 132.3 Mha in 1997. The cropland area affected by water erosion (moderate +) also decreased over the 15-year period. The area affected by water erosion was 40.6 Mha in 1982, 36.1 Mha in 1987, 28.7 Mha in 1992 and 25.5 Mha in 1997. In contrast to the cultivated cropland, the area under noncultivated cropland increased over the 10-year period from a lowest of 17.7 Mha in 1987 to 20.3 Mha in 1997. Nonetheless, the noncultivated area affected by water erosion remained the same, about 1.3 Mha.

53

**Plate 5.1**   Unprotected tilled fields develop surface seal and crust.

**Plate 5.2**   Clean cultivation leads to decline in soil structure and increase in soil erodibility.

**Plate 5.3**  Concentrated runoff causes transport of sediments and sediment-borne pollutants into the natural waters.

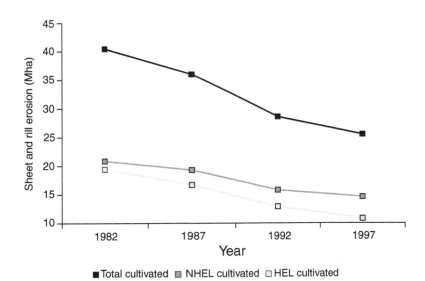

**Figure 5.1**  Land area affected by severe (>T) sheet and rill erosion from cultivated cropland in the U.S. (Refer to Table 5.1 for details.)

Table 5.1  Cultivated and Noncultivated Cropland Area Affected by Water Erosion on Private U.S. Land

| Land Use | Year | Total Area | Area Affected by Water Erosion | | | | | |
| | | | Light | Moderate | Severe | Extreme | Total (Moderate+) | % of Cropland |
| | | | Mha | | | | | |
| Cultivated | 1982 | 152.41 | 110.99 | 21.50 | 8.23 | 10.83 | 40.56 | 26.6 |
| | 1987 | 146.95 | 109.98 | 19.60 | 7.12 | 9.39 | 36.11 | 24.6 |
| | 1992 | 135.72 | 106.18 | 16.66 | 5.23 | 6.85 | 28.74 | 21.2 |
| | 1997 | 132.30 | 106.00 | 15.79 | 4.16 | 5.55 | 25.50 | 19.3 |
| Noncultivated | 1982 | 18.02 | 16.66 | 0.40 | 0.70 | 0.23 | 1.33 | 7.4 |
| | 1987 | 17.68 | 16.30 | 0.42 | 0.59 | 0.28 | 1.29 | 7.3 |
| | 1992 | 19.06 | 17.76 | 0.37 | 0.57 | 0.23 | 1.17 | 6.1 |
| | 1997 | 20.33 | 19.01 | 0.46 | 0.57 | 0.26 | 1.29 | 6.3 |

Total erosion area comprises sum of moderate, severe, and extreme forms, and % of the total is based on the total cropland area under that category. The erosion rate for cultivated land may be different if CRP land is factored out.

Source: From NRI, National Resources Inventory, Washington, D.C., 1997.

Therefore, the percentage of noncultivated cropland area affected by water erosion declined from 7.4% in 1982 to 6.3% in 1997 because marginal lands were converted to Conservation Reserve Program (CRP) (Table 5.1).

The data in Table 5.2 show the extent of erosion on cultivated vs. noncultivated, and nonirrigated (rainfed) vs. irrigated cropland. The nonirrigated cropland area decreased between 1982 (128.8 Mha) and 1992 (112.5 Mha), as did the land area affected by soil erosion from 38.1 Mha (29.6%) in 1982 to 26.4 Mha (23.5%) in 1992. About 10% of the irrigated cropland area was affected by erosion over this period. In comparison, 5 to 8% of the nonirrigated noncultivated area and 2% of the noncultivated irrigated area were affected by erosion for the period from 1982 to 1992 (Table 5.2).

## 5.2   SEVERITY OF SHEET AND RILL EROSION ON U.S. CROPLAND

The severity of sheet and rill erosion is assessed in terms of the soil erosion rate measured as a multiple of soil loss tolerance (T), the maximum permissible soil loss with little adverse impact on productivity and environment quality (Chapter 4). The qualitative categories of soil erosion (e.g., light, moderate, severe, and extreme) shown in Table 5.1 and Table 5.2 are changed into more quantitative classes of severity of soil erosion (based on T values) in the data in Table 5.3 to Table 5.7, Table 5.9 to Table 5.11, Table 5.13, and Table 5.14. Further, the cropland area is classified into highly erodible land (HEL) and nonhighly erodible (NHEL) depending on the susceptibility to erosion (Uri and Lewis, 1999; Uri, 2000). The data in Table 5.3 show different levels of sheet and rill erosion by water for the total cultivated and total noncultivated areas subdivided into HEL and NHEL categories. Total cultivated land area with susceptibility to erosion of >T was 40.5 Mha (26.6%) in 1982, 36.0 Mha (24.5%) in 1987, 28.6 Mha (21.1%) in 1992, and 25.5 Mha (19.2%) in 1997. In accordance with this reduction in the extent of erosion on total cultivated area, there was a corresponding decrease in land area affected by erosion on both NHEL cultivated and HEL cultivated categories.

In comparison to the total cultivated area, the total noncultivated cropland area was rather small, and was 18.0 Mha in 1982, 17.7 Mha in 1987, 19.1 Mha in 1992, and 20.3 Mha in 1997. The corresponding area affected by severe erosion (>T) was 0.69 Mha (3.8%) in 1982, 0.76 Mha (4.3%) in 1987, 0.66 Mha (3.5%) in 1992, and 0.77 Mha (3.8%) in 1997 (Table 5.3).

The data in Table 5.4 are the sum of cultivated and noncultivated cropland areas shown in Table 5.3 into total cropland classified into HEL and NHEL categories. The data in Table 5.4 are similar to that in Table 5.1, except that the severity of erosion is expressed quantitatively in terms of the T value. Total cropland area susceptible to severe erosion (>T) was 41.2 Mha (24.1%) in 1982, 36.9 Mha (36.9%) in 1987, 29.4 Mha (19.0%) in 1992, and 26.2 Mha (17.2%) in 1997. The area of total HEL cropland susceptible to erosion was 21.6 Mha (42.9%) in 1982, 20.0 Mha (42.3%) in 1987, 16.4 Mha (38.5%) in 1992, and 15.4 Mha (36.6%) in 1997. There was also a corresponding decline in the total NHEL cropland area susceptible to severe erosion (>T). The area was 19.5 Mha (16.3%) in 1982, 16.8 Mha (14.3%) in 1987, 12.9 Mha (11.5%) in 1992, and 10.8 Mha (9.8%) in 1997 (Table 5.4).

Table 5.2  Irrigated and Nonirrigated Cropland Area Affected by Water Erosion on U.S. Private Land (1992 Assessment)

| Land Use | Type | Year | Total Area | Area Affected by Water Erosion | | | | | |
| --- | --- | --- | --- | --- | --- | --- | --- | --- | --- |
| | | | | Light | Moderate | Severe | Extreme | Total (Moderate+) | % of Total |
| | | | | Mha | | | | | |
| Cultivated | Nonirrigated | 1982 | 128.76 | 90.15 | 19.73 | 7.93 | 10.42 | 38.086 | 29.6 |
| | | 1987 | 122.87 | 88.88 | 17.81 | 6.82 | 8.84 | 33.47 | 27.2 |
| | | 1992 | 112.46 | 85.61 | 15.04 | 4.94 | 6.40 | 26.38 | 23.5 |
| | Irrigated | 1982 | 19.50 | 17.26 | 1.34 | 0.27 | 0.30 | 1.91 | 9.8 |
| | | 1987 | 19.20 | 16.96 | 1.34 | 0.27 | 0.29 | 1.90 | 9.9 |
| | | 1992 | 19.31 | 17.11 | 1.31 | 0.27 | 0.28 | 1.86 | 9.6 |
| Noncultivated | Nonirrigated | 1982 | 16.67 | 15.01 | 0.79 | 0.17 | 0.43 | 1.39 | 8.3 |
| | | 1987 | 16.97 | 15.03 | 0.89 | 0.17 | 0.62 | 0.79 | 4.7 |
| | | 1992 | 17.15 | 15.68 | 0.68 | 0.12 | 0.43 | 1.23 | 7.2 |
| | Irrigated | 1982 | 5.50 | 4.95 | 0.037 | 0.006 | 0.043 | 0.086 | 1.6 |
| | | 1987 | 5.60 | 5.05 | 0.036 | 0.007 | 0.050 | 0.093 | 1.7 |
| | | 1992 | 5.87 | 5.30 | 0.041 | 0.005 | 0.049 | 0.095 | 1.6 |

Total erosion area comprises sum of moderate, severe, and extreme, and the % of the total is based on the total cropland area under that category.

Table 5.3  Severity of Sheet and Rill Erosion on NHEL and HEL Cropland

| Land Use | Year | Total Area | ≤T | 1–2T | 2–3T | 3–4T | 4–5T | >5T | Total Area >T | >T as % of the Total Area |
|---|---|---|---|---|---|---|---|---|---|---|
| | | | | | Mha | | | | | |
| 1. Total cultivated | 1982 | 152.4 | 111.9 | 21.4 | 7.5 | 3.7 | 2.3 | 5.6 | 40.5 | 26.6 |
| | 1987 | 146.9 | 110.9 | 19.5 | 6.7 | 3.4 | 2.0 | 4.4 | 36.0 | 24.5 |
| | 1992 | 135.7 | 107.0 | 16.6 | 5.5 | 2.5 | 1.3 | 2.7 | 28.6 | 21.1 |
| | 1997 | 132.3 | 106.9 | 15.7 | 4.8 | 2.1 | 1.0 | 1.9 | 25.5 | 19.2 |
| NHEL cultivated | 1982 | 106.5 | 89.5 | 15.7 | 3.3 | 0.46 | 0.026 | 0.003 | 19.489 | 18.3 |
| | 1987 | 106.5 | 89.7 | 13.9 | 2.5 | 0.31 | 0.018 | 0.002 | 16.730 | 15.7 |
| | 1992 | 100.7 | 87.8 | 11.1 | 1.6 | 0.16 | 0.017 | 0.0003 | 12.877 | 12.8 |
| | 1997 | 98.5 | 87.7 | 9.6 | 1.1 | 0.11 | 0.014 | 0.0003 | 10.824 | 11.0 |
| HEL cultivated | 1982 | 43.4 | 22.4 | 5.7 | 4.2 | 3.2 | 2.2 | 5.6 | 20.9 | 48.2 |
| | 1987 | 40.5 | 21.2 | 5.6 | 4.2 | 3.1 | 2.0 | 4.4 | 19.3 | 47.7 |
| | 1992 | 35.0 | 19.2 | 5.5 | 3.9 | 2.4 | 1.3 | 2.7 | 15.8 | 45.1 |
| | 1997 | 33.8 | 17.5 | 6.1 | 3.7 | 1.9 | 1.0 | 1.9 | 14.6 | 43.2 |
| | | | | | 10³ ha | | | | | |
| 2. Total noncultivated | 1982 | 18018 | 17328 | 395 | 136 | 62 | 28 | 69 | 690 | 3.8 |
| | 1987 | 17682 | 16926 | 417 | 147 | 71 | 33 | 87 | 755 | 4.3 |
| | 1992 | 19059 | 18398 | 370 | 141 | 63 | 24 | 62 | 660 | 3.5 |
| | 1997 | 20330 | 19559 | 458 | 153 | 54 | 31 | 75 | 771 | 3.8 |
| NHEL noncultivated | 1982 | 10970 | 10903 | 55 | 8 | 4 | 0 | 0.04 | 67 | 0.6 |
| | 1987 | 10873 | 10803 | 57 | 9 | 4 | 0.1 | 0.04 | 70 | 0.6 |
| | 1992 | 11529 | 11458 | 59 | 8 | 4 | 0.5 | 0 | 72 | 0.6 |
| | 1997 | 12017 | 11949 | 62 | 4 | 1.1 | 1.1 | 0 | 68 | 0.6 |
| HEL noncultivated | 1982 | 7048 | 6425 | 340 | 128 | 58 | 29 | 69 | 624 | 8.9 |
| | 1987 | 6809 | 6123 | 360 | 138 | 67 | 87 | 87 | 685 | 10.1 |
| | 1992 | 7530 | 6940 | 312 | 133 | 59 | 62 | 62 | 589 | 7.8 |
| | 1997 | 8313 | 7610 | 396 | 149 | 53 | 75 | 75 | 703 | 8.5 |

NHEL = nonhighly erodible land; HEL = highly erodible land.

**Table 5.4  Severity of Sheet and Rill Erosion in U.S. Total, HEL and NHEL Cropland**

| Land Use | Year | Total Area | ≤T | 1–2T | 2–3T | 3–4T | 4–5T | >5T | Total Area >T | >T as % of the Total Area |
|---|---|---|---|---|---|---|---|---|---|---|
| | | | | | | Mha | | | | |
| Total cropland | 1982 | 170.4 | 129.3 | 21.8 | 7.7 | 3.7 | 2.3 | 5.7 | 41.2 | 24.1 |
| | 1987 | 164.6 | 127.8 | 20.0 | 6.8 | 3.5 | 2.1 | 4.5 | 36.9 | 22.4 |
| | 1992 | 154.8 | 125.5 | 17.0 | 5.7 | 2.6 | 1.3 | 2.8 | 29.4 | 19.0 |
| | 1997 | 152.6 | 126.4 | 16.2 | 4.9 | 2.1 | 1.0 | 2.0 | 26.2 | 17.2 |
| Total HEL (cropland) | 1982 | 50.4 | 28.8 | 6.0 | 4.3 | 3.3 | 2.3 | 5.7 | 21.6 | 42.9 |
| | 1987 | 47.3 | 27.3 | 6.0 | 4.3 | 3.2 | 2.0 | 4.5 | 20.0 | 42.3 |
| | 1992 | 42.6 | 26.2 | 5.8 | 4.0 | 2.4 | 1.4 | 2.8 | 16.4 | 38.5 |
| | 1997 | 42.1 | 26.8 | 6.5 | 3.9 | 2.0 | 1.0 | 2.0 | 15.4 | 36.6 |
| Total NHEL (cropland) | 1982 | 120.0 | 100.4 | 15.7 | 3.3 | 0.5 | 0.03 | 0.003 | 19.53 | 16.3 |
| | 1987 | 117.3 | 100.5 | 14.0 | 2.5 | 0.3 | 0.02 | 0.002 | 16.82 | 14.3 |
| | 1992 | 112.2 | 99.3 | 11.1 | 1.6 | 0.16 | 0.02 | 0.0003 | 12.90 | 11.5 |
| | 1997 | 110.5 | 99.7 | 9.6 | 1.1 | 0.11 | 0.01 | 0.0003 | 10.80 | 9.8 |

## 5.3  MAGNITUDE OF SOIL EROSION ON U.S. CROPLAND

While the possible extent of erosion susceptibility is measured in land area affected by different severity (T value) of erosion, the magnitude of erosion is expressed in terms of the amount of soil displaced as computed by the Revised Universal Soil Loss Equation (RUSLE) in Mg of soil. The data in Table 5.5 are similar to those in Table 5.3, except that the land area is replaced by the magnitude of soil erosion measured in the units of million metric tons (MMT) or $10^6$ Mg under different severity (multiples of T) classes. The magnitude of soil displaced by sheet and rill erosion from total U.S. cropland was 1553 MMT (1.53 billion metric tonnes) in 1982, 1355 MMT in 1987, 1079 MMT in 1992, and 959 MMT in 1997 (Figure 5.2). As expected, the drastic decline in erosion over this period occurred in the HEL cultivated cropland (Plate 5.4 and Plate 5.5) from 826 MMT in 1982 to 440 MMT in 1997. In comparison, the decline in the magnitude of erosion from 1982 to 1997 was 679 MMT to 488 MMT in NHEL-cultivated cropland. This notable decline in the magnitude of erosion was due to conversion of HEL to CRP and adoption of best management practices (BMPs) on NHEL.

In contrast to the cultivated land, the magnitude of erosion on noncultivated cropland generally increased over the 15-year period. The magnitude of erosion from total noncultivated cropland was 28.3 MMT in 1982, 29.0 MMT in 1987, 27.5 MMT in 1992, and 30.3 MMT in 1997. The magnitude of erosion in noncultivated cropland increased (Table 5.5), and the relative increase was more in HEL- than NHEL-noncultivated land.

The data in Table 5.6 are the sum of cultivated and noncultivated cropland into total cropland. The magnitude of sheet and rill erosion on total HEL and NHEL cropland, respectively, was 847 and 687 MMT in 1982, 732 and 623 MMT in 1987, 542 and 537 MMT in 1992, and 463 and 496 MMT in 1997. The corresponding rate of decline in the magnitude of erosion from total NHEL cropland was 12.7 MMT/yr (Table 5.6).

## 5.4  RATE OF SOIL EROSION ON U.S. CROPLAND

The rate of water erosion processes from U.S. cropland, expressed as Mg/ha/yr, is shown in Table 5.7, and in Figure 5.3 and Figure 5.4. The rate of soil erosion decreased over the 15-year period in all cropland categories. In cultivated cropland, the erosion rate was 8.1 Mg/ha/yr in 1982, 7.8 Mg/ha/yr in 1987, 6.0 Mg/ha/yr in 1992, and 5.6 Mg/ha/yr in 1997. As one would expect, the highest rate of erosion was observed on HEL-cultivated land. The rate was 16.6 Mg/ha/yr in 1982, 16.1 Mg/ha/yr in 1987, 13.9 Mg/ha/yr in 1992, and 12.1 Mg/ha/yr in 1997 (Table 5.7). In contrast to the cultivated HEL cropland with a range of 12.1 to 16.6 Mg/ha/yr, the rate was drastically lower in noncultivated HEL cropland with a range of only 0.9 to 1.8 Mg/ha/yr.

The data in Table 5.8 are a summary of the details presented in Table 5.5 and Table 5.7. Uri and Lewis (1999) reported that total erosion on U.S. cropland decreased from 1.7 billion Mg in 1982 to 1.0 billion Mg in 1997, a reduction of

Table 5.5  The Magnitude of Sheet and Rill Erosion from NHEL and HEL Cropland under Different Severity Classes

| Land Use | Year | ≤T | 1–2T | 2–3T | 3–4T | 4–5T | >5T | Total |
|---|---|---|---|---|---|---|---|---|
| | | | | | $10^6$ Mg | | | |
| **1. Total cultivated** | 1982 | 466.0 | 290.4 | 168.5 | 115.1 | 89.2 | 403.9 | 1,533.1 |
| | 1987 | 445.9 | 264.5 | 150.3 | 106.1 | 79.9 | 307.8 | 1,354.5 |
| | 1992 | 420.0 | 224.0 | 124.1 | 78.1 | 51.5 | 181.7 | 1,079.4 |
| | 1997 | 414.2 | 212.3 | 107.9 | 62.6 | 38.6 | 123.0 | 958.6 |
| NHEL cultivated | 1982 | 375.6 | 212.8 | 75.1 | 14.0 | 1.0 | 0.09 | 678.6 |
| | 1987 | 361.3 | 187.4 | 55.9 | 9.7 | 0.8 | 0.05 | 615.2 |
| | 1992 | 341.4 | 146.9 | 36.0 | 4.7 | 0.7 | 0.02 | 529.7 |
| | 1997 | 334.1 | 126.1 | 23.8 | 3.6 | 0.5 | 0.02 | 488.1 |
| HEL cultivated | 1982 | 75.4 | 73.3 | 90.8 | 99.5 | 87.2 | 400.0 | 826.2 |
| | 1987 | 70.4 | 73.0 | 91.5 | 94.8 | 78.0 | 302.9 | 710.6 |
| | 1992 | 63.2 | 73.2 | 85.7 | 71.7 | 50.1 | 178.1 | 522.1 |
| | 1997 | 63.2 | 81.5 | 81.4 | 57.7 | 37.1 | 119.4 | 440.3 |
| **2. Total noncultivated** | 1982 | 15.0 | 4.3 | 2.6 | 1.6 | 1.0 | 3.8 | 28.3 |
| | 1987 | 14.2 | 4.4 | 2.8 | 1.6 | 1.1 | 4.9 | 29.0 |
| | 1992 | 15.4 | 3.8 | 2.5 | 1.6 | 0.7 | 3.5 | 27.5 |
| | 1997 | 16.9 | 4.8 | 2.7 | 1.4 | 0.9 | 3.6 | 30.3 |
| NHEL noncultivated | 1982 | 7.00 | 0.64 | 0.21 | 0.10 | 0 | 0 | 7.95 |
| | 1987 | 6.65 | 0.62 | 0.19 | 0.12 | 0.005 | 0 | 7.59 |
| | 1992 | 6.71 | 0.65 | 0.17 | 0.12 | 0.02 | 0 | 7.67 |
| | 1997 | 7.12 | 0.71 | 0.09 | 0.02 | 0.04 | 0 | 7.98 |
| HEL noncultivated | 1982 | 8.00 | 3.68 | 2.42 | 1.53 | 0.97 | 3.80 | 20.30 |
| | 1987 | 7.58 | 3.76 | 2.62 | 1.56 | 1.08 | 4.90 | 21.50 |
| | 1992 | 8.65 | 3.18 | 2.33 | 1.50 | 0.65 | 3.50 | 19.81 |
| | 1997 | 9.75 | 4.05 | 2.58 | 1.34 | 0.89 | 3.56 | 22.16 |

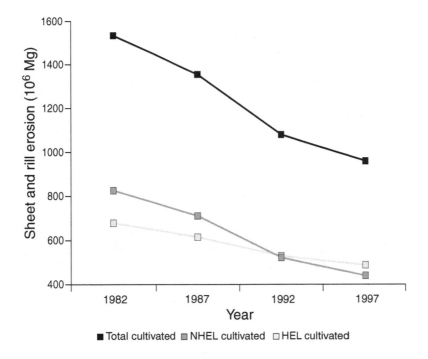

**Figure 5.2** The magnitude of sheet and rill erosion from cultivated cropland in the U.S. A large fraction of erosion is contributed by HEL cultivated land prior to conversion to CRP.

**Plate 5.4** An example of HEL cropland.

**Plate 5.5**  Plowing and other practices leading to soil disturbance and lack of protective vegetal cover can exacerbate erosion even on prime lands.

**Table 5.6   Magnitude of Sheet and Rill Erosion on Total HEL and NHEL Cropland**

| Land Use | Year | ≤T | 1–2T | 2–3T | 3–4T | 4–5T | >5T | Total |
|---|---|---|---|---|---|---|---|---|
| | | | | | $10^6$ Mg | | | |
| Total HEL | 1982 | 83.4 | 76.9 | 93.2 | 101.0 | 88.2 | 403.8 | 846.5 |
| (cropland) | 1987 | 78.0 | 76.7 | 94.2 | 96.3 | 79.1 | 307.8 | 732.1 |
| | 1992 | 71.9 | 76.4 | 88.0 | 73.2 | 50.7 | 181.7 | 541.9 |
| | 1997 | 73.0 | 85.5 | 84.4 | 59.0 | 38.1 | 122.9 | 462.9 |
| Total NHEL | 1982 | 382.5 | 213.4 | 75.3 | 14.2 | 1.0 | 0.09 | 686.5 |
| (cropland) | 1987 | 367.9 | 188.0 | 56.1 | 9.9 | 0.8 | 0.05 | 622.8 |
| | 1992 | 348.1 | 147.6 | 36.0 | 4.9 | 0.8 | 0.02 | 537.4 |
| | 1997 | 341.2 | 126.8 | 23.9 | 3.7 | 0.6 | 0.02 | 496.2 |

39.2%. The corresponding rate declined from 9.9 Mg/ha/yr in 1982 to 6.4 Mg/ha/yr in 1997, with a reduction of 35.4% (Table 5.8). Between 1995 and 1997, however, there was a slight increase in both the magnitude and rate of erosion. The magnitude of erosion increased from 0.94 billion Mg to 1.01 billion Mg, and the corresponding rate of erosion increased from 5.9 Mg/ha/yr to 6.4 Mg/ha/yr (Table 5.8).

## 5.5   PRIME CROPLAND AND SEVERITY OF SHEET AND RILL EROSION

Cropland is comprised of prime land and nonprime land. The prime land has few or no limitations to crop production. Because of the gentle terrain, prime land has slight or no risk of soil erosion. The area of prime cropland was 93.5 Mha in

Table 5.7   Rate of Soil Erosion by Water on HEL and NHEL Cropland

| Land Use | Year | <T | 1–2T | 2–3T | 3–4T | 4–5T | >5T |
|---|---|---|---|---|---|---|---|
| | | | | Mg soil/ha/yr | | | |
| **1. Total cultivated** | 1982 | 1.6 | 15.0 | 25.5 | 35.8 | 45.0 | 93.6 |
| | 1987 | 1.6 | 15.2 | 25.8 | 36.1 | 44.6 | 88.0 |
| | 1992 | 1.3 | 15.0 | 25.8 | 35.6 | 43.9 | 87.8 |
| | 1997 | 1.3 | 15.0 | 25.8 | 35.6 | 44.6 | 90.0 |
| NHEL cultivated | 1982 | 1.6 | 15.5 | 26.4 | 37.6 | 46.8 | 64.1 |
| | 1987 | 1.6 | 15.5 | 26.2 | 37.6 | 47.0 | 64.3 |
| | 1992 | 1.3 | 15.2 | 26.4 | 37.6 | 45.9 | 58.5 |
| | 1997 | 1.3 | 15.5 | 26.4 | 37.0 | 48.2 | 62.5 |
| HEL cultivated | 1982 | 1.3 | 14.6 | 24.4 | 34.3 | 44.1 | 95.4 |
| | 1987 | 1.3 | 14.8 | 24.9 | 34.5 | 43.0 | 89.8 |
| | 1992 | 1.6 | 14.6 | 25.1 | 34.0 | 42.8 | 88.9 |
| | 1997 | 1.6 | 14.6 | 24.4 | 34.3 | 44.1 | 92.2 |
| **2. Total noncultivated** | 1982 | 0.2 | 14.1 | 23.7 | 36.5 | 45.7 | 108.9 |
| | 1987 | 0.2 | 14.6 | 23.5 | 33.4 | 39.0 | 99.0 |
| | 1992 | 0.2 | 15.2 | 23.5 | 31.8 | 43.5 | 83.3 |
| | 1997 | 0.2 | 14.6 | 24.6 | 23.7 | 45.0 | 84.2 |
| NHEL noncultivated | 1982 | 0 | 15.2 | 28.2 | 37.2 | 49.3 | 75.7 |
| | 1987 | 0.2 | 15.0 | 27.1 | 37.0 | 46.8 | 56.4 |
| | 1992 | 0 | 15.7 | 25.1 | 30.9 | 46.4 | 71.2 |
| | 1997 | 0 | 15.2 | 26.9 | 38.8 | 48.6 | 76.8 |
| HEL noncultivated | 1982 | 0.2 | 13.4 | 21.7 | 36.5 | 44.6 | 109.8 |
| | 1987 | 0.2 | 14.1 | 22.6 | 32.5 | 37.0 | 100.4 |
| | 1992 | 0.2 | 14.8 | 23.3 | 32.5 | 39.2 | 83.8 |
| | 1997 | 0.2 | 14.1 | 23.1 | 20.8 | 44.8 | 84.4 |

1982, 91.4 Mha in 1987, 87.4 Mha in 1992, and 86.0 Mha in 1997 (Table 5.9). The area of prime cropland decreased over the 15-year period between 1982 and 1997. The erosion risk (>T) of prime cropland was on 21.3% of the total prime land area in 1982, 18.6% in 1987, 15.4% in 1992, and 3.5% in 1997. As the data in Table 5.9 show, most erosion on prime land occurred only on the cultivated cropland.

There is also nonprime cropland used for cultivation. Total nonprime cropland area was 76.9 Mha in 1982, 73.3 Mha in 1987, 67.3 Mha in 1992, and 66.6 Mha in 1997. Similar to the prime cropland, the area under nonprime cropland also decreased over the 15-year period at an average rate of 0.7 Mha/yr or 0.9%/yr. The erosion risk on nonprime cropland, as one would expect, was more than that on prime cropland. The area of nonprime cropland susceptible to severe erosion (>T) also decreased over the 15-year period, and was 21.3 Mha (27.6% of the total) in 1982, 19.6 Mha (26.7%) in 1987, 15.8 Mha (23.5%) in 1992, and 14.5 Mha (21.8%) in 1997 (Table 5.9; Figure 5.5).

The magnitude of soil erosion on prime cropland for different severity classes (multiple of T) is shown in Table 5.10. Similar to the land area, the magnitude of soil erosion on total and cultivated prime cropland decreased over the 15-year period (Figure 5.6). Because most of the erosion on prime land occurred on cultivated areas,

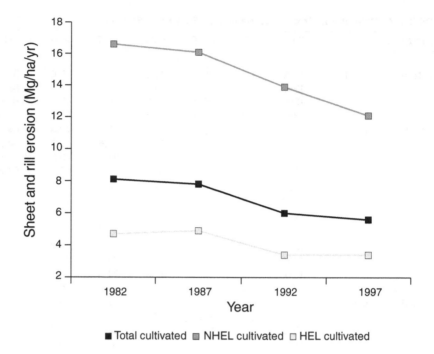

Total cultivated ■ NHEL cultivated □ HEL cultivated

**Figure 5.3**   The rate of sheet and rill erosion from total, NHEL- and HEL-cultivated cropland in the U.S.

the mean rate of decline in the magnitude of erosion on prime cultivated land was similar to that on the total prime land.

The magnitude of soil erosion was greater on the nonprime cropland vs. the prime cropland, and was 864 MMT in 1982, 758 MMT in 1987, 571 MMT in 1992, and 496 MMT in 1997 (Figure 5.7). The rate of decline in the magnitude of erosion over the 15-year period was 24.5 MMT/yr or 2.8%/yr. Similar to the erosion on the total prime cropland, most of the erosion occurred on cultivated nonprime cropland. Therefore, the rate of decline in the magnitude of erosion on nonprime cultivated land was also 24.8 MMT/yr.

The rate of soil erosion by water on prime cropland and nonprime cropland is shown by the data in Table 5.11. The rate of soil erosion on total nonprime cropland is more than double that of the total prime cropland for each year. The ratio of the rate of erosion on total nonprime to total prime cropland was 2.4 in 1982, 2.3 in 1987, 2.4 in 1992, and 2.3 in 1997. The rate of soil erosion on total prime cropland and total nonprime cropland declined over the 15-year period.

Similar to rate on total land, the rate of soil erosion on nonprime cultivated land was much higher than that on the prime cultivated land (Figure 5.8). The ratio of the rate of erosion on nonprime cultivated land to prime cultivated land was 2.7 in 1982, 2.6 in 1987, 2.8 in 1992, and 2.4 in 1997 (Figure 5.9). The rate of soil erosion on all croplands decreased over the 15-year period from 1982 to 1997.

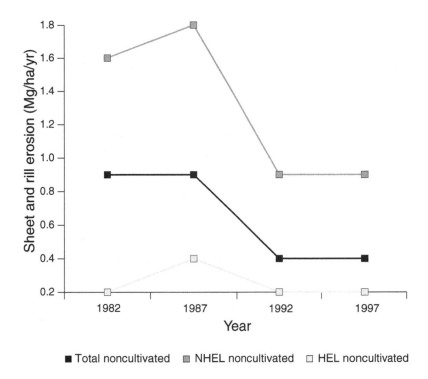

Sheet and rill erosion (Mg/ha/yr)

Year

■ Total noncultivated    ▨ NHEL noncultivated    □ HEL noncultivated

**Figure 5.4** The rate of sheet and rill erosion from total, NHEL- and HEL-noncultivated cropland in the U.S.

**Table 5.8 Temporal Changes in Sheet and Rill Erosion on U.S. Cropland**

| Year | Total Erosion (billion Mg) | Soil Erosion Rate (Mg/ha/yr) |
|------|---------------------------|------------------------------|
| 1982 | 1.69 | 9.92 |
| 1987 | 1.41 | 8.31 |
| 1992 | 1.08 | 7.03 |
| 1995 | 0.94 | 5.91 |
| 1996 | 0.98 | 6.07 |
| 1997 | 1.01 | 6.41 |

*Source*: Recalculated from Uri, N.D. and Lewis, J.A., *J. Sust. Agric.*, 14, 63-82, 1999.

## 5.6 EROSION RATE ON HIGHLY ERODIBLE CROPLAND

Uri (2000) reported the extent of water erosion on different types of U.S. cropland. The values for area under cropland of 170.5 Mha for 1982 and 154.8 Mha for 1992 shown in Table 5.12 are similar to those shown in Table 3.5. The average rate of erosion on all cropland was 9.2 Mg/ha/yr in 1982 and 6.9 Mg/ha/yr in 1992, a reduction of 25% over the 10-year period. There was a corresponding decrease in soil erosion rate on all categories of cropland. The soil erosion rate for 1982 and

Table 5.9 Distribution of Prime Cropland in Relation to Severity of Sheet and Rill Erosion

| Land Use | Year | Total Area | ≤T | 1–2T | 2–3T | 3–4T | 4–5T | >5T | Total Area >T | >T as % of the Total Area |
|---|---|---|---|---|---|---|---|---|---|---|
| | | | | | | **Mha** | | | | |
| **1. Total prime cropland** | 1982 | 93.5 | 73.6 | 13.3 | 3.8 | 1.3 | 0.7 | 0.9 | 20.0 | 21.3 |
| | 1987 | 91.4 | 74.2 | 11.7 | 3.1 | 1.1 | 0.5 | 0.6 | 17.0 | 18.6 |
| | 1992 | 87.4 | 73.9 | 9.6 | 2.4 | 0.8 | 0.3 | 0.4 | 13.5 | 15.4 |
| | 1997 | 86.0 | 74.3 | 8.7 | 1.9 | 0.6 | 0.2 | 0.2 | 11.6 | 13.5 |
| (i) Prime cultivated | 1982 | 86.9 | 67.1 | 13.1 | 3.8 | 1.4 | 0.7 | 0.9 | 19.9 | 23.0 |
| | 1987 | 84.8 | 67.8 | 11.6 | 3.1 | 1.1 | 0.5 | 0.6 | 16.9 | 19.9 |
| | 1992 | 80.4 | 67.0 | 9.5 | 2.4 | 0.8 | 0.3 | 0.4 | 13.4 | 16.8 |
| | 1997 | 78.5 | 67.0 | 8.6 | 1.9 | 0.6 | 0.2 | 0.24 | 11.5 | 14.7 |
| | | | | | | **10³ ha** | | | | |
| (ii) Prime noncultivated | 1982 | 6616 | 6515 | 74 | 14 | 4.5 | 3.0 | 5.2 | 100.7 | 1.5 |
| | 1987 | 6509 | 6398 | 75 | 20 | 6.1 | 1.4 | 7.6 | 110.1 | 1.7 |
| | 1992 | 7034 | 6922 | 76 | 23 | 4.9 | 2.0 | 5.5 | 111.4 | 1.6 |
| | 1997 | 7466 | 7361 | 74 | 21 | 2.3 | 2.0 | 4.6 | 103.9 | 1.4 |
| | | | | | | **Mha** | | | | |
| **2. Total nonprime cropland** | 1982 | 76.9 | 55.7 | 8.6 | 3.9 | 2.4 | 1.6 | 4.8 | 21.3 | 27.6 |
| | 1987 | 73.3 | 53.6 | 8.2 | 3.7 | 2.3 | 1.5 | 3.9 | 19.6 | 26.7 |
| | 1992 | 67.3 | 51.5 | 7.3 | 3.3 | 1.8 | 1.0 | 2.4 | 15.8 | 23.5 |
| | 1997 | 66.6 | 52.1 | 7.4 | 3.1 | 1.5 | 0.8 | 1.7 | 14.5 | 21.8 |
| (i) Nonprime cultivated | 1982 | 65.5 | 44.9 | 8.2 | 3.8 | 2.3 | 1.6 | 4.7 | 28.8 | 44.0 |
| | 1987 | 62.1 | 43.0 | 7.9 | 3.6 | 2.3 | 1.5 | 3.8 | 21.4 | 34.4 |
| | 1992 | 55.3 | 40.0 | 7.0 | 3.1 | 1.7 | 1.0 | 2.3 | 18.2 | 32.9 |
| | 1997 | 53.8 | 39.9 | 7.1 | 2.9 | 1.5 | 0.8 | 1.7 | 14.0 | 26.0 |
| (ii) Nonprime noncultivated | 1982 | 11.4 | 10.8 | 0.32 | 0.12 | 0.06 | 0.03 | 0.06 | 0.59 | 5.2 |
| | 1987 | 11.2 | 10.5 | 0.34 | 0.13 | 0.06 | 0.03 | 0.08 | 0.64 | 5.7 |
| | 1992 | 12.0 | 11.5 | 0.29 | 0.12 | 0.06 | 0.02 | 0.06 | 0.55 | 4.6 |
| | 1997 | 12.9 | 12.2 | 0.38 | 0.13 | 0.05 | 0.03 | 0.07 | 0.66 | 5.1 |

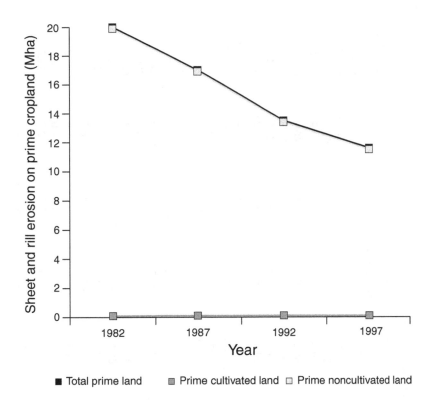

**Figure 5.5**  The prime land area affected by different severity of erosion.

1992, respectively, was 17.0 Mg/ha/yr vs. 13.0 Mg/ha/yr for highly erodible cropland, a reduction of 24%; 19.7 Mg/ha/yr vs. 15.5 Mg/ha/yr for highly erodible
cultivated cropland, a reduction of 21%; 4.5 Mg/ha/yr vs. 3.8 Mg/ha/yr for highly
erodible noncultivated cropland, a reduction of 16%; 5.8 Mg/ha/yr vs. 4.7 Mg/ha/yr
for nonhighly erodible cropland, a reduction of 19%; 6.3 Mg/ha/yr vs. 5.4 Mg/ha/yr
for nonhighly erodible cultivated cropland, a reduction of 14%; and 0.9 Mg/ha/yr
for nonhighly erodible noncultivated cropland for both years.

The data in Table 5.13 are based on the sum of all croplands (prime and
nonprime), but further subdivided into HEL and NHEL croplands. Similar to the
trends on prime cropland, the erosion rate was much higher on HEL than NHEL
cropland. The ratio of the rate of erosion on HEL to NHEL cropland was 3.3 in
1982, 3.1 in 1987, 3.7 in 1992, and 3.2 in 1997. The rate of erosion decreased over
the 15-year period in total for HEL and NHEL cropland (Figure 5.10).

## 5.7  DISTRIBUTION OF ERODED CROPLAND IN THE U.S.

The data in Table 5.14 show the distribution of water erosion by severity classes
for all states in the conterminous U.S. The areas with the highest severely eroded land
(with >4T value) occurred in Iowa (0.51 Mha), Missouri (0.29 Mha), Illinois (0.27 Mha),

Table 5.10  Magnitude of Erosion on Prime Cropland for Different Severity Classes

| Land Use | Year | <T | 1–2T | 2–3T | 3–4T | 4–5T | >5T | Total |
|---|---|---|---|---|---|---|---|---|
| | | | | | 10⁶ Mg soil | | | |
| **1. Total prime cropland** | 1982 | 297.7 | 177.7 | 82.1 | 40.9 | 24.8 | 46.4 | 669.6 |
| | 1987 | 287.0 | 155.7 | 66.8 | 33.8 | 18.8 | 34.5 | 596.6 |
| | 1992 | 275.5 | 126.5 | 51.4 | 22.6 | 11.9 | 20.2 | 508.1 |
| | 1997 | 271.7 | 112.9 | 40.0 | 16.4 | 7.4 | 12.9 | 461.3 |
| (i) Prime cultivated | 1982 | 292.9 | 176.8 | 81.8 | 40.8 | 24.7 | 46.1 | 663.1 |
| | 1987 | 282.3 | 154.9 | 66.4 | 33.6 | 18.8 | 34.2 | 590.2 |
| | 1992 | 270.7 | 125.6 | 51.0 | 22.4 | 11.8 | 20.0 | 501.5 |
| | 1997 | 266.5 | 112.1 | 39.7 | 16.4 | 7.4 | 12.7 | 454.8 |
| (ii) Prime noncultivated | 1982 | 4.7 | 0.90 | 0.28 | 20.13 | 0.10 | 0.29 | 6.40 |
| | 1987 | 4.6 | 0.87 | 0.41 | 0.18 | 0.06 | 0.34 | 6.46 |
| | 1992 | 4.8 | 0.87 | 0.44 | 0.15 | 0.06 | 0.28 | 6.60 |
| | 1997 | 5.2 | 0.82 | 0.37 | 0.05 | 0.07 | 0.20 | 6.71 |
| **2. Total nonprime cropland** | 1982 | 168.3 | 112.7 | 86.4 | 74.2 | 64.4 | 357.5 | 863.5 |
| | 1987 | 158.9 | 108.9 | 83.4 | 72.3 | 61.1 | 273.3 | 757.9 |
| | 1992 | 144.4 | 97.5 | 72.7 | 55.5 | 39.6 | 161.5 | 571.2 |
| | 1997 | 141.5 | 99.4 | 67.8 | 46.2 | 31.2 | 110.1 | 496.2 |
| (i) Nonprime cultivated | 1982 | 158.1 | 109.2 | 84.1 | 76.4 | 63.5 | 354.0 | 845.3 |
| | 1987 | 149.3 | 105.4 | 81.1 | 70.8 | 60.1 | 268.7 | 735.4 |
| | 1992 | 133.8 | 94.5 | 70.7 | 54.0 | 39.0 | 158.2 | 550.2 |
| | 1997 | 130.7 | 95.4 | 65.5 | 44.9 | 30.3 | 106.7 | 473.5 |
| (ii) Nonprime noncultivated | 1982 | 10.2 | 3.4 | 2.3 | 1.5 | 0.9 | 3.5 | 21.8 |
| | 1987 | 9.6 | 3.5 | 2.4 | 1.5 | 1.0 | 4.6 | 22.6 |
| | 1992 | 10.6 | 3.0 | 2.1 | 1.5 | 0.6 | 3.2 | 21.0 |
| | 1997 | 11.7 | 3.9 | 2.3 | 1.3 | 0.9 | 3.4 | 23.5 |

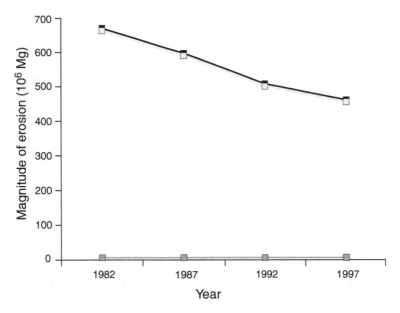

**Figure 5.6** The magnitude of soil erosion on prime cropland.

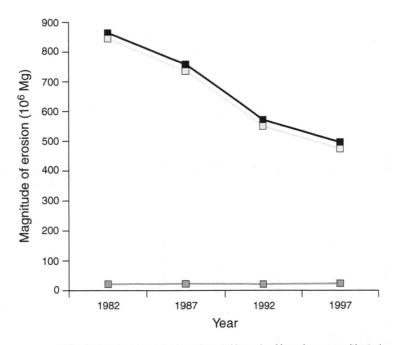

**Figure 5.7** The magnitude of soil erosion on nonprime cropland.

Table 5.11  Temporal Changes in Soil Erosion Rate on Prime Cropland in the U.S.

| Land Use | Year | ≤T | 1–2T | 2–3T | 3–4T | 4–5T | >5T | Average Rate |
|---|---|---|---|---|---|---|---|---|
| | | | | | Mg/ha/yr | | | |
| **1. Total prime cropland** | 1982 | 1.1 | 15.7 | 26.9 | 37.4 | 47.0 | 77.1 | 4.5 |
| | 1987 | 1.1 | 15.9 | 26.4 | 37.9 | 47.7 | 75.0 | 4.5 |
| | 1992 | 1.1 | 15.5 | 26.8 | 37.4 | 45.5 | 71.7 | 3.4 |
| | 1997 | 0.9 | 15.5 | 26.9 | 37.0 | 46.6 | 69.7 | 3.1 |
| Prime cultivated | 1982 | 1.3 | 15.7 | 26.9 | 37.4 | 46.8 | 75.0 | 4.7 |
| | 1987 | 1.3 | 15.7 | 26.7 | 37.9 | 47.7 | 73.0 | 4.7 |
| | 1992 | 1.1 | 15.5 | 26.9 | 37.4 | 45.5 | 71.5 | 3.6 |
| | 1997 | 1.1 | 15.5 | 26.9 | 37.0 | 46.6 | 69.4 | 3.6 |
| Prime noncultivated | 1982 | 0.2 | 14.8 | 26.4 | 39.9 | 50.8 | 175.8 | 0.7 |
| | 1987 | 0.2 | 16.6 | 25.8 | 35.2 | 38.3 | 172.9 | 0.7 |
| | 1992 | 0 | 16.6 | 23.5 | 36.1 | 46.4 | 85.6 | 0.2 |
| | 1997 | 0 | 14.8 | 24.9 | 38.5 | 47.7 | 85.8 | 0.2 |
| **2. Total nonprime cropland** | 1982 | 1.6 | 14.8 | 24.6 | 34.7 | 44.1 | 96.5 | 10.8 |
| | 1987 | 1.6 | 14.8 | 25.0 | 34.7 | 42.8 | 90.3 | 10.5 |
| | 1992 | 1.3 | 14.8 | 24.9 | 34.3 | 43.2 | 90.2 | 8.1 |
| | 1997 | 1.3 | 14.8 | 24.6 | 34.3 | 43.5 | 93.6 | 7.2 |
| Nonprime cultivated | 1982 | 1.8 | 14.8 | 24.6 | 34.7 | 44.1 | 96.5 | 12.5 |
| | 1987 | 1.8 | 14.8 | 25.0 | 34.7 | 42.8 | 90.5 | 12.1 |
| | 1992 | 1.8 | 14.8 | 24.9 | 34.3 | 43.2 | 90.5 | 9.9 |
| | 1997 | 1.6 | 14.8 | 24.6 | 34.5 | 43.5 | 93.9 | 8.7 |
| Nonprime noncultivated | 1982 | 0.2 | 13.9 | 22.4 | 35.8 | 43.9 | 90.0 | 0.9 |
| | 1987 | 0.2 | 13.7 | 23.1 | 35.8 | 43.9 | 90.0 | 0.9 |
| | 1992 | 0.2 | 14.8 | 23.5 | 29.8 | 43.2 | 83.1 | 0.7 |
| | 1997 | 0.2 | 14.3 | 24.4 | 20.4 | 44.4 | 83.8 | 0.4 |

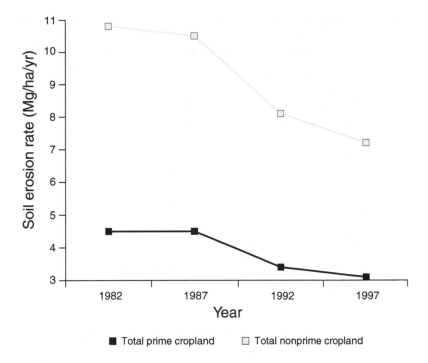

**Figure 5.8**   The rate of soil erosion on total prime and nonprime cropland.

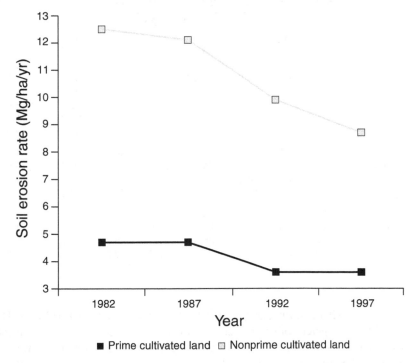

**Figure 5.9**   The rate of soil erosion on prime cultivated and nonprime cultivated cropland.

Table 5.12   Average Sheet and Rill Erosion in U.S. Cropland Assessed in 1982 and 1992

| Year | Land Use | Area (Mha) | Water Erosion (Mg/ha/yr) |
|---|---|---|---|
| 1982 | Cropland (excluding CRP) | 170.5 | 9.2 |
| | Highly erodible cropland | 50.6 | 17.0 |
| | Cultivated | 41.9 | 19.7 |
| | Noncultivated | 8.7 | 4.5 |
| | Nonhighly erodible cropland | 99.6 | 5.8 |
| | Cultivated | 106.4 | 6.3 |
| | Noncultivated | 13.4 | 0.9 |
| 1992 | Cropland (excluding CRP) | 154.8 | 6.9 |
| | Highly erodible cropland | 42.7 | 13.0 |
| | Cultivated | 33.6 | 15.5 |
| | Noncultivated | 9.1 | 3.8 |
| | Nonhighly erodible land | 112.1 | 4.7 |
| | Cultivated | 98.2 | 5.4 |
| | Noncultivated | 13.9 | 0.9 |

Source: Modified from Uri, N.D., J. Sust. Agric., 16, 71-94, 2000.

Table 5.13   Rate of Soil Erosion on U.S. Cropland for Highly Erodible Land and Nonhighly Erodible Land

| Land Use | Year | ≤T | 1–2T | 2–3T | 3–4T | 4–5T | >5T | Average Rate |
|---|---|---|---|---|---|---|---|---|
| | | | | | Mg/ha/yr | | | |
| Total cropland | 1982 | 1.3 | 15.0 | 25.5 | 35.8 | 45.0 | 93.6 | 7.4 |
| | 1987 | 1.3 | 15.2 | 25.8 | 36.1 | 44.4 | 88.3 | 7.2 |
| | 1992 | 1.3 | 15.0 | 25.8 | 35.6 | 43.9 | 87.8 | 5.4 |
| | 1997 | 1.1 | 15.0 | 25.8 | 35.6 | 50.5 | 89.8 | 4.9 |
| | | | | | | | 95.6 | 14.3 |
| | | | | | | | 89.8 | 13.9 |
| | | | | | | | 88.9 | 11.4 |
| | | | | | | | 92.3 | 9.9 |
| Total HEL (cropland) | 1982 | 1.1 | 14.6 | 24.4 | 34.3 | 44.1 | 95.6 | 14.3 |
| | 1987 | 1.1 | 14.8 | 24.9 | 34.5 | 43.0 | 89.8 | 13.9 |
| | 1992 | 1.1 | 14.6 | 25.1 | 34.0 | 42.8 | 88.9 | 11.4 |
| | 1997 | 1.1 | 14.6 | 24.4 | 34.0 | 42.6 | 92.3 | 9.9 |
| Total NHEL (cropland) | 1982 | 1.3 | 15.5 | 26.4 | 37.6 | 46.8 | 64.1 | 4.3 |
| | 1987 | 1.3 | 15.5 | 26.2 | 37.6 | 47.0 | 64.1 | 4.5 |
| | 1992 | 1.3 | 15.2 | 26.4 | 37.4 | 45.9 | 58.7 | 3.1 |
| | 1997 | 1.1 | 15.5 | 26.4 | 37.0 | 48.2 | 62.5 | 3.1 |

Tennessee (0.17 Mha), North Carolina (0.12 Mha), and Kentucky (0.10 Mha). These lands need to be targeted for erosion control measures and conversion to restorative land uses. The data in Table 5.15 show the magnitude of soil erosion by water for each state in the conterminous U.S. The total soil erosion of 1.84 billion Mg was

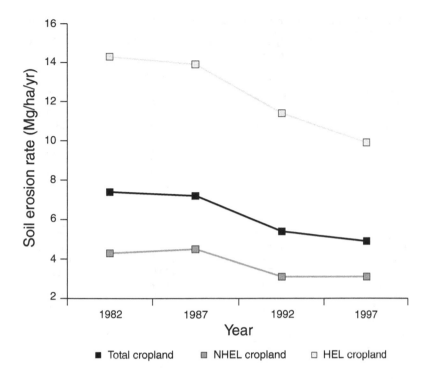

**Figure 5.10**  Temporal changes in rate of erosion from HEL and NHEL.

comprised of 1.04 billion Mg from cultivated cropland, 0.05 billion Mg from non-cultivated cropland, 0.114 billion Mg from pastureland, 0.44 billion Mg from range-land, and 0.21 billion Mg from other lands.

## 5.8  CONCLUSIONS

A drastic reduction in soil erosion on U.S cropland (e.g., in area affected, the magnitude of soil moved, and the rate of erosion) during the last two decades of the 20th century is a success story (Friend, 1992). This reduction is attributed to the adoption of recommended management practices (Plate 5.6), conversion of HEL, and adoption of other improved technologies. The data in Table 5.16 show that soil erosion on cropland in the U.S. during the 1930s was progressing at a disastrous rate. Although survey techniques and terminology used during the 1930s were different than those during the 1990s, the comparison highlights the importance of adopting judicious land use and BMPs to reduce the risks of soil erosion, and enhance soil resilience to recover soil processes. Bennett (1939) reported that half of the U.S. land area had been affected by erosion before or during the 1930s, and vast areas

Table 5.14    U.S. Cropland Subjected to Different Severity of Water Erosion

| State | Slightly Eroded (<T) | Moderately Eroded (1–4 T) | Severely Eroded (>4 T) |
|-------|---------------------|---------------------------|------------------------|
|       | 1000 ha | | |
| Alabama | 675.7 | 556.7 | 41.7 |
| Arizona | 484.3 | 0.5 | 0 |
| Arkansas | 2661.1 | 461.3 | 7.2 |
| California | 3970.4 | 70.2 | 28.9 |
| Colorado | 3297.5 | 306.4 | 15.6 |
| Connecticut | 77.0 | 12.4 | 3.1 |
| Delaware | 186.4 | 15.7 | 0 |
| Florida | 1137.3 | 74.7 | 1.5 |
| Georgia | 1361.6 | 672.9 | 59.8 |
| Idaho | 1806.2 | 446.4 | 14.2 |
| Illinois | 7577.9 | 1905.7 | 273.3 |
| Indiana | 4557.4 | 830.6 | 82.7 |
| Iowa | 7238.7 | 2364.0 | 513.8 |
| Kansas | 9948.2 | 768.3 | 38.4 |
| Kentucky | 1486.7 | 472.5 | 102.3 |
| Louisiana | 1931.4 | 480.9 | 5.4 |
| Maine | 169.0 | 12.2 | 0 |
| Maryland | 518.1 | 135.8 | 23.4 |
| Massachusetts | 101.7 | 7.4 | 1.1 |
| Michigan | 3346.2 | 266.3 | 25.2 |
| Minnesota | 7817.7 | 757.9 | 70.4 |
| Mississippi | 1529.0 | 716.3 | 72.9 |
| Missouri | 3902.9 | 1210.4 | 289.9 |
| Montana | 5673.8 | 393.4 | 19.7 |
| Nebraska | 6540.2 | 1038.9 | 21.0 |
| Nevada | 308.6 | 0 | 0 |
| New Hampshire | 55.4 | 1.5 | 0.4 |
| New Jersey | 196.1 | 57.5 | 9.5 |
| New Mexico | 763.1 | 2.8 | 0 |
| New York | 1976.7 | 268.7 | 28.3 |
| North Carolina | 1754.0 | 543.3 | 115.5 |
| North Dakota | 9539.9 | 471.5 | 6.6 |
| Ohio | 4027.4 | 747.0 | 55.0 |
| Oklahoma | 3477.0 | 595.0 | 9.2 |
| Oregon | 1225.8 | 285.7 | 17.0 |
| Pennsylvania | 1688.8 | 501.3 | 75.4 |
| Rhode Island | 8.2 | 1.9 | 0 |
| South Carolina | 1029.9 | 167.8 | 9.8 |
| South Dakota | 6101.2 | 529.6 | 23.6 |
| Tennessee | 1232.2 | 566.0 | 168.1 |
| Texas | 10192.6 | 1224.7 | 24.6 |
| Utah | 714.2 | 16.0 | 4.6 |
| Vermont | 238.9 | 16.7 | 1.4 |
| Virginia | 893.2 | 230.5 | 50.9 |
| Washington | 1904.4 | 768.3 | 58.1 |
| West Virginia | 340.3 | 26.3 | 3.8 |
| Wisconsin | 3618.6 | 675.4 | 83.8 |
| Wyoming | 911.7 | 8.0 | 0 |
| Total | 130194.0 | 21685.8 | 2646.2 |
|       | (± 733.5) | (± 292.1) | (±93.2) |

Source: From NRI, National Resources Inventory, NRCS, Washington, D.C., 1992.

Table 5.15 Magnitude of Water Erosion in Relation to Land Use for Different States in the Conterminous U.S. (1000 Mg/yr)

| State | Cropland | | Forestland | Pastureland | Rangeland | Other | Total |
| --- | --- | --- | --- | --- | --- | --- | --- |
| | Cultivated | Noncultivated | | | | | |
| Alabama | 17547.3 | 165.7 | 0.0 | 1777.9 | 153.5 | 4102.8 | 23747.2 |
| Arizona | 560.9 | 44.6 | 0.0 | 7.4 | 24119.5 | 1022.6 | 25755.1 |
| Arkansas | 23889.2 | 172.4 | 0.0 | 6415.1 | 283.4 | 1303.5 | 32063.6 |
| California | 5890.5 | 1477.4 | 0.0 | 176.7 | 41152.7 | 2740.3 | 51437.7 |
| Colorado | 16819.3 | 307.5 | 0.0 | 343.4 | 30012.3 | 4887.5 | 52370.1 |
| Connecticut | 508.6 | 185.3 | 0.0 | 16.7 | 0.0 | 283.8 | 994.2 |
| Delaware | 920.9 | 4.4 | 0.0 | 12.1 | 0.0 | 39.8 | 977.1 |
| Florida | 2732.8 | 436.0 | 0.0 | 552.5 | 237.5 | 2060.1 | 6018.8 |
| Georgia | 23472.3 | 242.0 | 0.0 | 1009.2 | 0.0 | 2479.8 | 27203.2 |
| Idaho | 14670.8 | 475.6 | 0.0 | 467.6 | 3932.8 | 1874.5 | 21421.4 |
| Illinois | 92747.3 | 1785.7 | 0.0 | 2535.6 | 0.0 | 5707.1 | 102775.7 |
| Indiana | 39430.0 | 1027.6 | 0.0 | 1443.1 | 0.0 | 2407.7 | 44308.5 |
| Iowa | 118991.9 | 2980.1 | 0.0 | 4372.5 | 0.0 | 4177.3 | 130521.8 |
| Kansas | 51364.3 | 585.5 | 0.0 | 1504.0 | 12644.7 | 2564.7 | 68663.2 |
| Kentucky | 22401.4 | 2131.8 | 0.0 | 16955.7 | 0.0 | 36898.2 | 78387.1 |
| Louisiana | 18948.4 | 119.8 | 0.0 | 364.2 | 15.6 | 415.9 | 19864.0 |
| Maine | 453.7 | 79.4 | 0.0 | 23.2 | 0.0 | 319.3 | 875.6 |
| Maryland | 6435.3 | 422.3 | 0.0 | 520.9 | 0.0 | 674.5 | 8053.1 |
| Massachusetts | 275.2 | 37.2 | 0.0 | 26.4 | 0.0 | 117.2 | 456.0 |
| Michigan | 14337.8 | 1162.8 | 0.0 | 524.5 | 0.0 | 1568.3 | 17593.3 |
| Minnesota | 40441.8 | 1599.5 | 0.0 | 886.6 | 0.0 | 1297.3 | 44225.2 |
| Mississippi | 27960.6 | 396.4 | 0.0 | 4575.3 | 0.0 | 4637.3 | 37569.7 |
| Missouri | 64796.1 | 2292.1 | 0.0 | 18115.5 | 353.8 | 8515.6 | 94073.3 |
| Montana | 21972.8 | 507.1 | 0.0 | 531.4 | 28452.7 | 1520.1 | 52984.1 |
| Nebraska | 56915.7 | 1036.9 | 0.0 | 1430.7 | 17743.7 | 1848.2 | 78975.2 |
| Nevada | 33.4 | 13.5 | 0.0 | 12.9 | 4007.5 | 139.3 | 4206.4 |
| New Hampshire | 67.0 | 39.1 | 0.0 | 38.3 | 0.0 | 130.6 | 275.1 |
| New Jersey | 2474.0 | 200.5 | 0.0 | 69.3 | 0.0 | 1102.2 | 3846.1 |

(continued)

Table 5.15 Magnitude of Water Erosion in Relation to Land Use for Different States in the Conterminous U.S. (1000 Mg/yr) (*Continued*)

| State | Cropland | | Forestland | Pastureland | Rangeland | Other | Total |
|---|---|---|---|---|---|---|---|
| | Cultivated | Noncultivated | | | | | |
| New Mexico | 1248.6 | 61.6 | 0.0 | 31.0 | 27266.2 | 785.1 | 29392.6 |
| New York | 9014.9 | 3571.1 | 0.0 | 950.4 | 0.0 | 3634.5 | 17170.8 |
| North Carolina | 28040.0 | 533.3 | 0.0 | 1760.4 | 0.0 | 2790.7 | 33124.5 |
| North Dakota | 29894.6 | 780.3 | 0.0 | 537.3 | 14454.2 | 1259.6 | 46926.1 |
| Ohio | 30694.3 | 3009.4 | 0.0 | 3486.3 | 0.0 | 6526.9 | 43716.9 |
| Oklahoma | 25752.9 | 238.1 | 0.0 | 4898.3 | 18764.6 | 3598.8 | 53252.7 |
| Oregon | 9327.9 | 441.4 | 0.0 | 885.3 | 29133.5 | 993.4 | 40781.5 |
| Pennsylvania | 16990.0 | 4776.2 | 0.0 | 2083.1 | 0.0 | 28891.0 | 52740.3 |
| Rhode Island | 32.6 | 23.4 | 0.0 | 3.6 | 0.0 | 36.4 | 96.0 |
| South Carolina | 8355.3 | 212.6 | 0.0 | 458.4 | 0.0 | 1705.6 | 10732.0 |
| South Dakota | 29023.6 | 979.2 | 0.0 | 483.5 | 22125.7 | 14408.5 | 67020.6 |
| Tennessee | 29966.5 | 1205.9 | 0.0 | 3239.6 | 0.0 | 5457.4 | 39869.4 |
| Texas | 64693.3 | 540.9 | 0.0 | 8096.6 | 88294.7 | 8462.7 | 170088.2 |
| Utah | 1278.9 | 170.1 | 0.0 | 72.8 | 19509.9 | 1250.6 | 22282.2 |
| Vermont | 446.1 | 310.4 | 0.0 | 43.0 | 0.0 | 507.0 | 1306.5 |
| Virginia | 10354.4 | 1672.4 | 0.0 | 11127.0 | 0.0 | 1698.1 | 24843.9 |
| Washington | 26015.0 | 555.2 | 0.0 | 404.0 | 4659.7 | 5728.5 | 37362.3 |
| West Virginia | 945.2 | 569.8 | 0.0 | 9193.7 | 0.0 | 20768.0 | 31476.7 |
| Wisconsin | 26152.2 | 5630.3 | 0.0 | 1497.4 | 0.0 | 2247.0 | 35527.0 |
| Wyoming | 1259.4 | 262.7 | 0.0 | 315.5 | 49984.8 | 1244.0 | 53066.4 |
| Total | 1036544.9 | 45464.3 | 0.0 | 114276.4 | 437303.1 | 206829.4 | 1840418.5 |

**Table 5.16   Estimated Extent of Soil Erosion in the U.S. during the 1930s**

| Erosion Type | Land Area Affected (Mha) |
|---|---|
| Total U.S. land area (48 states) | 771 |
| **Degraded land of all types** | |
| • Essentially ruined | 24 |
| • Severely eroded | 90 |
| • Moderately eroded in the beginning | 310 |
| **Damaged cropland** | |
| • Essentially ruined for cultivation | 20 |
| • Severely damaged | 20 |
| • One-half to all topsoil gone | 40 |
| • Moderately eroded, erosion beginning | 40 |
| **Land not currently damaged (forest, swamp, or marsh)** | 284 |
| **Land with unidentified damage** | 59 |

*Source*: Recalculated from Bennett, H.H., *Soil Conservation*, McGraw-Hill, New York, 1939.

**Plate 5.6**   Conversion from plow till to conservation tillage reduces risks of soil erosion.

of cropland have essentially been ruined (or irreversibly degraded). Further, 75% of the U.S. cropland during the 1930s was affected by some degree of erosion. In fact, lack of increase in yield of corn and cotton during the 60-year period from 1871 to 1930 was attributed to the adverse effect of severe erosion (Bennett, 1939). Yet, estimates of soil erosion made 60 years later as reported in this chapter show that the extent, magnitude, and rate of soil erosion on cropland were less in the 1990s than in the 1930s. It is likely that risks of soil erosion were exaggerated (Miller et al., 1985).

It is encouraging to note that conversion of HEL to other uses (e.g., forest) and adoption of BMPs on cropland (no till, mulch farming, growing cover crops) increased production while decreasing the rate of soil erosion. In fact, much of the so-called "ruined land" described by Bennett in the 1930s was used for production of timber and forages in the 1980s and 1990s. The resilience characteristics of these soils, through conversion to an appropriate land use and adoption of BMPs, transformed these "ruined lands" of the 1930s into productive and environmentally safe lands of the 1990s. It was the high resilience (Lal, 1997) of these lands that brought about the classic transformation.

The data in Table 5.17 and Table 5.18 are regression equations showing the rate of decline of specific parameters over the 15-year period between 1982 and 1997. These data demonstrate that soil erosion can be managed through conversion to an appropriate land use (e.g., CRP) and adoption of BMPs. Conversion of HEL to CRP was very effective in reducing the rate of soil erosion. Regression equations in Table 5.17 show that the rate of decline of erosion (extent, magnitude, and rate) was high on HEL cropland. The regression equations in Table 5.18 show similar trends with regard to decline in soil erosion hazard on prime and nonprime cultivated land.

**Table 5.17   Regression Equation between Sheet and Rill Erosion on Cultivated Cropland with Time (Years) between 1982 (Base Year) and 1997 (15 Years)**

| Dependent Variable (Y) | Regression Equation | $R^2$ |
|---|---|---|
| I. Extent of erosion (Mha) | | |
|    (i) Total cultivated cropland | Y = 40.5 − 1.05x | 0.978 |
|    (ii) NHEL-cultivated cropland | Y = 19.5 − 0.598x | 0.989 |
|    (iii) HEL-cultivated | Y = 21.0 − 0.448x | 0.965 |
| II. Magnitude of erosion ($10^6$ Mg) | | |
|    (i) Total cultivated cropland | Y = 1531 − 40.0x | 0.98 |
|    (ii) NHEL-cultivated cropland | Y = 677 − 13.2x | 0.985 |
|    (iii) HEL-cultivated cropland | Y = 827 − 26.9x | 0.979 |
| III. Rate of erosion (Mg/ha/yr) | | |
|    (i) Total cultivated cropland | Y = 8.27 − 0.186x | 0.911 |
|    (ii) NHEL-cultivated cropland | Y = 4.91 − 0.108x | 0.736 |
|    (iii) HEL-cultivated cropland | Y = 17.0 − 0.34x | 0.950 |
| IV. Rate of erosion (Mg/ha/yr) | | |
|    (i) Total noncultivated cropland | Y = 0.95 − 0.04x | 0.80 |
|    (ii) NHEL-noncultivated cropland | Y = 0.28 − 0.004x | 0.067 |
|    (iii) HEL-noncultivated cropland | Y = 1.75 − 0.06x | 0.682 |

X = years after 1982. For example, the value of x for 1987 is 5 years and 1997 is 15 years.

**Table 5.18 Regression Equations between Sheet and Rill Erosion on Prime and Nonprime Cultivated and Noncultivated Cropland with Time between 1982 (Base Year) and 1997 (15 Years)**

| Dependent Variable (Y) | Regression Equation | $R^2$ |
|---|---|---|
| I. Extent of erosion (Mha) | | |
| (i) Total prime land | Y = 19.8 − 0.574x | 0.987 |
| (ii) Prime cultivated land | Y = 19.7 − 0.574x | 0.987 |
| (iii) Prime noncultivated land | Y = 0.105 + 0.0x | 0 |
| II. Magnitude of erosion ($10^6$ Mg) | | |
| (i) Total prime land | Y = 666 − 14.3x | 0.987 |
| (ii) Prime cultivated land | Y = 659 − 14.3x | 0.986 |
| (iii) Prime uncultivated land | Y = 6.4 + 0.02x | 0.999 |
| III. Magnitude of erosion ($10^6$ Mg) | | |
| (i) Total nonprime cropland | Y = 866 − 25.8x | 0.975 |
| (ii) Nonprime cultivated land | Y = 846 − 26.0x | 0.977 |
| (iii) Nonprime noncultivated land | Y = 21.7 − 0.07x | 0.178 |
| IV. Soil erosion rate (Mg/ha/yr) | | |
| (i) Total prime cropland | Y = 4.67 − 0.106x | 0.874 |
| (ii) Total nonprime cropland | Y = 11.1 − 0.264x | 0.922 |
| (iii) Prime cultivated land | Y = 4.81 − 0.088x | 0.80 |
| (iv) Nonprime cultivated land | Y = 12.8 − 0.272x | 0.944 |
| (v) Total cropland | Y = 7.62 − 0.186x | 0.907 |
| (vi) HEL cropland | Y = 14.7 − 0.314x | 0.940 |
| (vii) Non-HEL cropland | Y = 4.5 − 0.1x | 0.731 |

X = years after 1982. For example, the value of x for 1987 is 5 years and for 1997 is 15 years.

# REFERENCES

Bennett, H.H. *Soil Conservation*. McGraw Hill, New York, 1939.

Friend, J.A. Achieving soil sustainability. *J. Soil Water Cons.* 47, 156–157, 1992.

Lal, R. Degradation and resilience of soils. *Phil. Trans. R. Soc. Lond. (Biology),* 997–1010, 1997.

Miller, F.P., W.D. Rasmussen, and L.D. Meyer. Historical perspective of soil erosion in the United States. In *Soil Erosion and Crop Productivity*. R.F. Follett and B.A. Stewart (Eds.), ASA-CSSA-SSSA, Madison, WI, 1985, pp. 23–48.

NRI. National Resources Inventory. NRCS, Washington, D.C., 1992.

NRI. National Resources Inventory. NRCS, Washington, D.C., 1997.

Uri, N.D. Agriculture and the environment — the problem of soil erosion. *J. Sust. Agric.* 16, 71–94, 2000.

Uri, N.D. and J.A. Lewis. Agriculture and the dynamics of soil erosion in the United States. *J. Sust. Agric.* 14, 63–82, 1999.

# CHAPTER 6

# Soil Erosion by Water from Grazing, CRP, and Minor Lands

Grazing land, as used in this volume, comprises pastureland, rangeland, and grazed woodlands. Rangeland is the uncultivated land with natural vegetation cover that provides the habitat for grazing and browsing animals (Holechek et al., 1989). Rangelands are disturbed by grazing, natural or managed fires, road traffic, and mining activities. Therefore, rangelands cover a wide spectrum of ecosystems including perennial and annual grasslands, desert shrub, and forest (Sobecki et al., 2001). Rangelands are characterized by the type of vegetation, which depends on the precipitation and soil type:

- Desert shrublands exist in arid regions with annual precipitation of less than 250 mm, and the dominant soils are Aridisols and Entisols.
- Grasslands occur in regions with annual precipitation of 250 to 900 mm and with predominant soils being Mollisols and others characterized by deep profile and loamy texture.
- Savanna woodlands occur in regions with annual precipitation of 250 to 500 mm, and the predominant soils are mainly Alfisols.
- Forested rangelands occur in regions with annual precipitation of more than 500 mm and with soilscape dominated by Alfisols, Ultisols, and Oxisols. These rangelands occur in subtropics and tropical ecosystems.
- Tundra rangelands occur on soils with permafrost (Cryosols) and in regions with annual precipitation of 250 to 500 mm. There is little or no water erosion problems on rangelands due to a continuous vegetative cover throughout the year (Lal, 2001a). However, tracks formed by grazing animals can lead to formation of gullies in dry areas (Plate 6.1A and Plate 6.1B).

In contrast to rangelands, however, soils under pastures are periodically disturbed by cultivation to introduce new forage species and enhance productivity. They also receive more precipitation. Because pastures receive more agronomic inputs than rangelands (e.g., irrigation, fertilizers, and manures), the stocking rate is often more on pastures than rangeland. If pastures are not grazed, the hay is periodically cut and carried away for stall feeding. Consequently, risks of soil erosion are much greater on pastureland than on rangeland if the lands are tilled.

**Plate 6.1A**  Rangelands suffer from both wind and water erosion.

**Plate 6.1B**  Cattle footpaths can lead to gullying of rangeland.

Regardless of the management, soil erosion risks on grazing land are less than on cropland. Similar to cropland, however, the magnitude of soil erosion risks on grazing land also depends on climatic erosivity parameters, soil erodibility, ground cover and its management, and crusting created by grazing and/or the interaction between climate/soil/management (Lal, 2001a). Burning accentuates the risks of soil erosion on grazing lands (Hester et al., 1997). In addition, the water infiltration rate may decrease with a reduction in ground cover due to intense grazing and heavy

Table 6.1    Pastureland Area Affected by Water Erosion on Private U.S. Land

| Year | Total Area | Area Affected by Water Erosion | | | | Total (Moderate +) | % of Total Area |
|------|------|------|------|------|------|------|------|
| | | Light | Moderate | Severe | Extreme | | |
| | | | | Mha | | | |
| 1982 | 53.44 | 47.94 | 2.10 | 0.23 | 1.64 | 3.97 | 7.4 |
| 1987 | 51.87 | 46.80 | 1.93 | 0.18 | 1.44 | 3.55 | 6.8 |
| 1992 | 51.03 | 46.05 | 1.94 | 0.15 | 1.42 | 3.51 | 6.9 |
| 1997 | 48.58 | 44.19 | 1.77 | 0.11 | 1.10 | 2.97 | 6.1 |

Total erosion area comprises sum of moderate, severe and extreme forms, and % of the total is based on the total cropland area under that category.

stocking rate (Lal, 2001b). In addition, concentrated cattle traffic can cause gully erosion, especially due to overgrazing and poor management.

## 6.1 PASTURELAND AREA AFFECTED BY WATER EROSION

Estimates of the land area affected by water erosion per 1997 NRI data are shown in Table 6.1. Similar to the cropland, the land area under pasture also decreased from 53.4 Mha in 1982 to 48.6 Mha in 1997, at an average rate of 0.6%/yr. Accordingly, the land area affected by erosion also decreased from about 4.0 Mha in 1982 (7.4%) to 3.0 Mha (6.1%) in 1997. Whereas the area of cultivated cropland affected by water erosion ranged from 19 to 27% for the 1997 survey, and 26 to 38% for the 1992 survey, the pastureland area affected by erosion ranged only between 6 and 7.5% (Table 6.1).

The data in Table 6.2 subdivide grazing lands into pastureland, rangeland, and grazed forestland. Each of these three land uses is further subdivided into grazed and nongrazed categories. The percentage of the land area affected by water erosion is in the order rangeland > pastureland > forestland. There is only slight erosion on grazed or nongrazed forestland. In contrast with cropland, there are no trends in temporal changes in land area affected by erosion, except in the light category. The land area affected by light erosion increased over the 10-year period in grazed pastureland and in grazed rangeland. However, land area affected by light erosion decreased over the decade ending in 1992 both in nongrazed pastureland and nongrazed rangeland, but there was no change in percent of the eroded land.

## 6.2 SEVERITY OF SOIL EROSION ON PASTURELAND

The data in Table 6.3 show the distribution of pastureland with regard to the severity of soil erosion. The pastureland area affected by severe erosion (>T) was about 4 Mha (7.4%) in 1982, 3.5 Mha (6.8%) both in 1987 and 1992, and 3.0 Mha (6.2%) in 1997 (Figure 6.1). The pastureland area most affected by erosion was in the 1–2T class followed by that in the 2–3T class. The data in Table 6.3 showing distribution of pastureland in different severity classes in relation to T value are similar to those in Table 6.1 in relation to light, moderate, severe, and extreme categories.

Table 6.2  Grazing Land Area Affected by Water Erosion on U.S. Private Land (1992 Assessment)

| Land Use | Type | Year | Total Area | Area Affected by Water Erosion | | | | | |
| | | | | Light | Moderate | Severe | Extreme | Total (Moderate+) | % of Total |
| | | | | Mha | | | | | |
| Pastureland | Grazed | 1982 | 40.16 | 35.85 | 1.69 | 0.19 | 1.36 | 3.24 | 8.1 |
| | | 1987 | 42.73 | 38.44 | 1.66 | 0.14 | 1.34 | 3.14 | 7.3 |
| | | 1992 | 45.78 | 41.14 | 1.81 | 0.14 | 1.49 | 3.44 | 7.5 |
| | Nongrazed | 1982 | 13.23 | 11.90 | 0.38 | 0.058 | 0.35 | 0.79 | 6.0 |
| | | 1987 | 8.93 | 8.01 | 0.24 | 0.045 | 0.17 | 0.46 | 5.1 |
| | | 1992 | 5.20 | 4.66 | 0.12 | 0.030 | 0.078 | 0.23 | 4.4 |
| Rangeland | Grazed | 1982 | 155.04 | 131.17 | 9.39 | 0.45 | 11.44 | 21.28 | 13.7 |
| | | 1987 | 155.43 | 132.00 | 9.36 | 0.41 | 11.01 | 20.78 | 13.3 |
| | | 1992 | 156.36 | 132.49 | 9.47 | 0.39 | 11.44 | 21.30 | 13.6 |
| | Nongrazed | 1982 | 10.51 | 8.67 | 0.39 | 0.051 | 0.78 | 1.22 | 11.6 |
| | | 1987 | 7.65 | 6.15 | 0.31 | 0.040 | 0.67 | 1.02 | 13.3 |
| | | 1992 | 5.16 | 3.95 | 0.27 | 0.020 | 0.54 | 0.83 | 6.0 |
| Forestland | Grazed | 1982 | 26.59 | 0 | 0 | 0 | 0 | 0 | 0 |
| | | 1987 | 25.72 | 0 | 0 | 0 | 0 | 0 | 0 |
| | | 1992 | 25.49 | 0 | 0 | 0 | 0 | 0 | 0 |
| | Nongrazed | 1982 | 133.07 | 0 | 0 | 0 | 0 | 0 | 0 |
| | | 1987 | 134.54 | 0 | 0 | 0 | 0 | 0 | 0 |
| | | 1992 | 134.41 | 0 | 0 | 0 | 0 | 0 | 0 |

Total erosion area comprises sum of moderate, severe, and extreme forms; percent of the total is based on the total cropland area under that category.

**Table 6.3  Distribution of Total Pastureland in Relation to Severity of Sheet and Rill Erosion**

| Year | Total Area | ≤T | 1–2T | 2–3T | 3–4T | 4–5T | >5T | Total Area >T | >T as % of Total Area |
|------|-----------|------|------|------|------|------|------|------|------|
| | | | | | Mha | | | | |
| 1982 | 53.4 | 49.5 | 2.1 | 0.83 | 0.42 | 0.20 | 0.41 | 3.97 | 7.4 |
| 1987 | 51.9 | 48.3 | 1.9 | 0.71 | 0.35 | 0.19 | 0.38 | 3.53 | 6.8 |
| 1992 | 51.0 | 47.5 | 1.9 | 0.70 | 0.36 | 0.19 | 0.33 | 3.48 | 6.8 |
| 1997 | 48.6 | 45.6 | 1.8 | 0.59 | 0.25 | 0.13 | 0.24 | 3.01 | 6.2 |

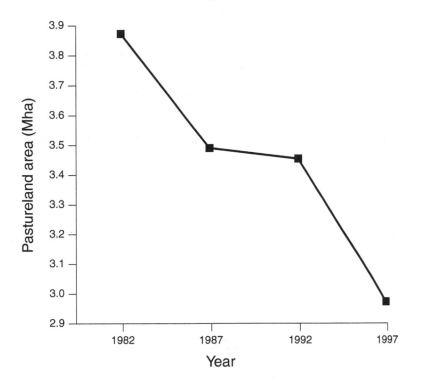

**Figure 6.1**  Temporal changes in total pastureland area affected by severe erosion.

The pastureland area affected by severe erosion (>2T) declined over the 15-year period between 1982 and 1997. The regression equations relating severity of erosion to time (between 1982 and 1997 taking 1982 as baseline) are shown in Equation 6.1 and Equation 6.2 for erosion severity ≥ light erosion and total area affected by erosion, respectively.

$$Y = 3.95 - 0.060x, \quad R^2 = 0.92 \qquad (6.1)$$

$$Y = 3.94 - 0.059x, \quad R^2 = 0.93 \qquad (6.2)$$

where x is the number of years after 1982. The rate of decline was 0.06 Mha/yr for total area affected by light + category and 0.059 Mha/yr for all categories of soil erosion.

Table 6.4　Magnitude of Water Erosion on Pastureland

| Year | ≤T | 1–2T | 2–3T | 3–4T | 4–5T | >5T | Total |
|------|------|------|------|------|------|------|-------|
| | | | | 10⁶ Mg | | | |
| 1982 | 54.8 | 22.0 | 14.7 | 9.8 | 5.7 | 20.3 | 127.3 |
| 1987 | 51.9 | 19.8 | 12.2 | 7.8 | 5.2 | 18.6 | 115.5 |
| 1992 | 51.0 | 19.7 | 11.9 | 7.9 | 5.4 | 15.8 | 111.7 |
| 1997 | 49.1 | 17.6 | 9.9 | 5.4 | 3.3 | 11.0 | 96.3 |

Table 6.5　Rate of Water Erosion on U.S. Pastureland

| Year | 1–2T | 2–3T | 3–4T | 4–5T | >5T |
|------|------|------|------|------|------|
| | | | Mg/ha/yr | | |
| 1982 | 14.3 | 22.6 | 32.5 | 48.4 | 73.0 |
| 1987 | 13.0 | 22.4 | 32.7 | 44.4 | 71.2 |
| 1992 | 11.9 | 21.7 | 27.6 | 44.1 | 71.9 |
| 1997 | 12.5 | 23.3 | 27.6 | 42.1 | 77.3 |

## 6.3 MAGNITUDE AND RATE OF SOIL EROSION ON PASTURELAND

The magnitude of soil erosion by water on U.S. pastureland, expressed as the quantity of soil displaced, is shown in Table 6.4. The total quantity of soil erosion was 127 million Mg in 1982, 116 million Mg in 1987, 112 million Mg in 1992, and 96 million Mg in 1997. The magnitude of soil erosion declined at an average rate of 2.1 million Mg/yr or 1.6%/yr. Similar to the severity of soil erosion, the rate of decline of erosion on pastureland was greater between 1982 and 1987, and 1992 and 1997 than between 1987 and 1992 (Table 6.5). However, the rate of erosion must be adjusted for the change in land area under that category. The magnitude of soil erosion by water was the greatest in 1–2T and >5T erosional classes.

## 6.4 SOIL EROSION ON CRP LAND

The potential area of Conservation Reserve Program (CRP) land that could be affected by water erosion increased between 1987 and 1997 because the land area under CRP progressively increased (Table 6.6). The land area under CRP was 5.6 Mha in 1987, 13.8 Mha in 1992, and 13.2 Mha in 1998. All other factors remaining the same, however, the area of CRP land affected by erosion decreases over time as the cover increases after grass was planted. Despite the increase in land area under CRP, the area affected by erosion decreased over this time. The CRP area affected by erosion (>T) was 0.72 Mha (12.9%) in 1987, 0.30 Mha (2.2%) in 1992, and 0.16 Mha (1.2%) in 1998. Establishment of grass vegetation cover and tree cover improved the soil structure, increased the infiltration rate, and minimized the risk of soil erosion by raindrop impact or flowing runoff.

Severity of sheet and rill erosion on CRP land in terms of the T value is shown in Table 6.7. The most severe erosion, as would be expected, occurred on the highly erodible land (HEL). The erosion-prone area on HEL-CRP was 0.52 Mha in 1987,

Table 6.6   CRP Land Area Affected by Water Erosion on Private U.S. Land

| Year | Total Area | Area Affected by Water Erosion | | | | Total (Moderate+) | % of Total Area |
|------|------------|-------|----------|--------|---------|-------------------|------------------|
| | | Light | Moderate | Severe | Extreme | | |
| | | | | Mha | | | |
| 1987 | 5.59 | 4.83 | 0.38 | 0.10 | 0.24 | 0.72 | 12.9 |
| 1992 | 13.78 | 13.28 | 0.17 | 0.04 | 0.09 | 0.30 | 2.2 |
| 1998 | 13.24 | 12.91 | 0.09 | 0.02 | 0.05 | 0.16 | 1.2 |

Total erosion area comprises sum of moderate, severe, and extreme forms, and % of the total is based on the total cropland area under that category.

0.24 Mha in 1992, and 0.05 Mha in 1997. The CRP land area with severe erosion (>T) was relatively less in the nonhighly erodible (NHEL)-CRP class.

The magnitude of soil erosion on CRP land, separated for HEL and NHEL categories, is shown in Table 6.8. Total soil erosion, expressed as the quantity of soil displaced, was 25.4 million Mg in 1987, 18.6 million Mg in 1992, and 11.2 million Mg in 1997. The magnitude of soil erosion from CRP land decreased between 1987 and 1997 even though the area under CRP increased over the same period. The average rate of decline in the magnitude of soil erosion over the 10-year period was 1.4 million Mg/yr although most of the decrease would be expected in the first and second years after cover was established.

Of the total magnitude, a large fraction of erosion on CRP land occurred on HEL. The fraction of the total erosion that occurred on HEL under CRP was 76.4% in 1987, 72.1% in 1992, and 7.3% in 1997. Most of the erosion in CRP-HEL category occurred within erosion classes of <T, 1–2T, and >5T. In contrast, erosion in NHEL-CRP occurred only in erosion classes of <3T (Table 6.8).

The mean rate of soil erosion on CRP land decreased drastically over the 10-year period (Table 6.9). The mean rate on all CRP land was 15.2 Mg/ha/yr in 1987, 1.6 Mg/ha/yr in 1982, and 0.7 Mg/ha/yr in 1997, although most changes may have occurred during the first year or two. The rate of erosion on HEL-CRP was several times more than that on NHEL-CRP. The ratio of soil erosion on HEL-CRP:NHEL-CRP was 4.1 in 1987, 5.5 in 1992, and 5.5 in 1997.

## 6.5 MINOR LAND AFFECTED BY WATER EROSION

Minor land encompasses miscellaneous land uses including urban land. The land area classed under the "minor land" category increased over the 10-year period from 1982 to 1992 (Table 6.10). The minor land area was 21.4 Mha in 1982, 21.6 Mha in 1987, and 22.1 Mha in 1992. While the area under minor land increased, the area affected by erosion slightly decreased over this time. The area affected by erosion was 1.74 Mha (8.1%) in 1982, 1.76 Mha (8.2%) in 1987, and 1.71 Mha (7.7%) in 1992 (Table 5.6). The minor land area affected by severe erosion decreased at an average rate of 3000 ha/yr between 1982 and 1992.

Table 6.7 Distribution of CRP Land in Relation to Severity of Sheet and Rill Erosion

| Land Use | Year | Total Area | ≤T | 1–2T | 2–3T | 3–4T | 4–5T | >5T | Total Area >T | >T as % of Total Area |
|---|---|---|---|---|---|---|---|---|---|---|
| | | Mha | | | | | 10³ ha | | | |
| Total CRP | 1987 | 5.56 | 4.86 | 380 | 130 | 70 | 40 | 10 | 720 | 12.9 |
| | 1992 | 13.78 | 13.49 | 170 | 40 | 30 | 20 | 40 | 300 | 2.2 |
| | 1997 | 13.23 | 13.09 | 90 | 20 | 11 | 16 | 11 | 150 | 1.1 |
| | | 10³ ha | | | | | | | | |
| CRP (HEL) | 1987 | 3443 | 2919 | 221 | 96 | 69 | 41 | 96 | 523 | 15.2 |
| | 1992 | 7797 | 7571 | 121 | 33 | 28 | 18 | 36 | 236 | 3.0 |
| | 1997 | 7754 | 7422 | 76 | 23 | 8 | 15 | 11 | 54 | 0.7 |
| CRP (NHEL) | 1987 | 2145 | 1944 | 163 | 34 | 4 | 0 | 0 | 201 | 9.4 |
| | 1992 | 5974 | 5918 | 47 | 8 | 0 | 1.5 | 0 | 57 | 1.0 |
| | 1997 | 5683 | 5668 | 9 | 2 | 3 | 0 | 0 | 14 | 0.2 |

Table 6.8 Severity of Water Erosion on CRP Land

| Land Use | Year | ≤T | 1–2T | 2–3T | 3–4T | 4–5T | >5T | Total >T |
|---|---|---|---|---|---|---|---|---|
| | | 10⁶ Mg | | | | | | |
| Total CRP | 1987 | 8.64 | 4.56 | 2.40 | 1.68 | 1.41 | 6.69 | 25.38 |
| | 1992 | 12.00 | 1.90 | 0.81 | 0.83 | 0.56 | 2.53 | 18.63 |
| | 1997 | 8.38 | 0.92 | 0.45 | 0.33 | 0.38 | 0.71 | 11.16 |
| CRP (HEL) | 1987 | 5.60 | 2.47 | 1.67 | 1.56 | 1.41 | 6.69 | 19.39 |
| | 1992 | 7.66 | 1.31 | 0.62 | 0.83 | 0.49 | 2.53 | 13.44 |
| | 1997 | 5.67 | 0.81 | 0.40 | 0.23 | 0.38 | 0.71 | 8.20 |
| CRP (NHEL) | 1987 | 3.04 | 2.09 | 0.73 | 0.12 | 0 | 0 | 5.98 |
| | 1992 | 4.34 | 0.59 | 0.18 | 0 | 0.07 | 0 | 5.18 |
| | 1997 | 2.70 | 0.11 | 0.05 | 0.10 | 0 | 0 | 2.96 |

Table 6.9 Water Erosion Rate on CRP Land

| Land Use | Year | ≤T | 1–2T | 2–3T | 3–4T | 4–5T | >5T |
|---|---|---|---|---|---|---|---|
| | | Mg/ha/yr | | | | | |
| Total CRP | 1987 | 1.3 | 14.8 | 25.3 | 34.7 | 44.4 | 94.5 |
| | 1992 | 0.4 | 15.0 | 24.2 | 32.0 | 38.5 | 90.0 |
| | 1997 | 0.2 | 15.5 | 24.9 | 29.1 | 31.1 | 110.0 |
| CRP (HEL) | 1987 | 1.1 | 14.8 | 26.0 | 34.7 | 44.8 | 95.6 |
| | 1992 | 0.4 | 15.0 | 24.0 | 32.0 | 37.0 | 90.0 |
| | 1997 | 0.2 | 15.5 | 22.8 | 27.8 | 22.8 | 110.0 |
| CRP (NHEL) | 1987 | 1.6 | 14.8 | 24.0 | 34.5 | 39.9 | 52.6 |
| | 1992 | 0.2 | 13.9 | 24.6 | 30.9 | 43.2 | N/A |
| | 1997 | 0 | 16.1 | 29.1 | 34.7 | 47.0 | N/A |

N/A = Data not available.

Table 6.10 Land Area Affected by Water Erosion on Minor Lands on Private U.S. Land (1992 Assessments)

| Year | Total Area | Area Affected by Water Erosion | | | | Total (Moderate+) | % of Total Area |
|---|---|---|---|---|---|---|---|
| | | Light | Moderate | Severe | Extreme | | |
| | | Mha | | | | | |
| 1982 | 21.39 | 7.86 | 0.45 | 0.49 | 0.80 | 1.74 | 8.1 |
| 1987 | 21.58 | 8.01 | 0.51 | 0.46 | 0.79 | 1.76 | 8.2 |
| 1992 | 22.10 | 8.57 | 0.18 | 0.40 | 0.83 | 1.71 | 7.7 |

Total erosion area comprises sum of moderate, severe, and extreme forms, and % of the total is based on the total cropland area under that category.

## 6.6 TOTAL LAND AREA AFFECTED BY WATER EROSION ON PRIVATE U.S. LAND

The extent of soil erosion on all classes of privately owned U.S. land is shown in Table 6.11 based on 1997 data. The area affected by soil erosion progressively declined over the 15-year period between 1982 and 1997. The total area affected by erosion (moderate, severe, and extreme categories) was 45.2 Mha (7.9%) in 1982, 41.15 Mha (7.2%) in 1987, 33.22 Mha (5.8%) in 1992, and 29.41 Mha (5.2%) in 1997. While the area affected by light erosion somewhat increased between 1982

Table 6.11   Total Land Area Affected by Water Erosion on Private U.S. Land

| Year | Total Area | Area Affected by Water Erosion | | | | Total (Moderate+) | % of Total Area |
|------|-----------|-------|----------|--------|---------|-------------------|-----------------|
|      |           | Light | Moderate | Severe | Extreme |                   |                 |
|      |           |       |   Mha    |        |         |                   |                 |
| 1982 | 575.99 | 175.58 | 24.00 | 8.53 | 12.70 | 45.23 | 7.9 |
| 1987 | 572.81 | 177.91 | 22.34 | 7.46 | 11.35 | 41.15 | 7.2 |
| 1992 | 569.08 | 183.27 | 19.15 | 5.47 | 8.60 | 33.22 | 5.8 |
| 1997 | 564.28 | 182.11 | 18.11 | 4.34 | 6.96 | 29.41 | 5.2 |

Total erosion area comprises sum of moderate, severe, and extreme forms, and % of the total is based on the total cropland area under that category.

Table 6.12   Total Land Area Affected by Water Erosion on Private U.S. Land (1992 Assessment)

| Year | Total Area | Area Affected by Water Erosion | | | | Total (Moderate+) | % of Total Area |
|------|-----------|-------|----------|--------|---------|-------------------|-----------------|
|      |           | Light | Moderate | Severe | Extreme |                   |                 |
|      |           |       |   Mha    |        |         |                   |                 |
| 1982 | 575.99 | 322.82 | 34.19 | 9.60 | 25.93 | 69.72 | 12.1 |
| 1987 | 572.81 | 323.36 | 32.52 | 8.49 | 24.03 | 65.04 | 11.4 |
| 1992 | 579.08 | 327.73 | 29.41 | 6.36 | 21.63 | 57.40 | 9.9 |

Total erosion area comprises sum of moderate, severe, and extreme forms, and % of the total is based on the total cropland area under that category.

and 1997, area affected by other three erosion classes (e.g., moderate, severe, and extreme) decreased drastically. Land area affected by moderate erosion decreased from 8.5 Mha in 1982 to 4.3 Mha in 1997, at an average rate of 0.3 Mha/yr. Land area affected by extreme erosion decreased from 12.7 Mha in 1982 to about 7.0 Mha in 1997, at an average rate of 0.4 Mha/yr or 3.0%/yr.

The data in Table 6.12 based on resource inventory conducted in 1992 show more land area affected by erosion than that shown in the survey conducted in 1997. Land area affected by water erosion on private U.S. land was 69.7 Mha (12.1%) in 1982, 65.0 Mha (11.4%) in 1987, and 57.4 Mha (9.9%) in 1997. Trends in land area affected by different severity classes were also similar to the data shown in Table 7.12. The land area affected by the light erosion category decreased and that affected by other three categories (moderate, severe, and extreme) increased over the decade ending in 1992. The land area affected by moderate erosion decreased from 34.2 Mha in 1982 to 29.4 Mha in 1992, at an average rate of about 0.5 Mha/yr or 1.4%/yr. The total area affected by severe erosion decreased from 9.6 Mha in 1982 to 6.4 Mha in 1992, at an average rate of 0.3 Mha/yr or 3.3%/yr. The total area affected by extreme erosion decreased from 25.9 Mha in 1982 to 21.6 Mha in 1992, at an average rate of 0.4 Mha/yr or 1.7%/yr (Table 6.12).

## 6.7 CONCLUSION

Soil erosion on grazing land is less severe than that on cropland. Similar to cropland, the magnitude/severity and rate of erosion on pastureland also decreased over the 15-year period between 1982 and 1997.

**Table 6.13  Regional Distribution of HEL Area**

| Production Region | HEL Area (Mha) |
|---|---|
| Pacific | 1.66 |
| Mountain | 8.87 |
| Northern Plains | 8.58 |
| Southern Plains | 7.53 |
| Lake States | 2.31 |
| Corn Belt | 9.39 |
| Delta States | 1.13 |
| Northeast | 2.96 |
| Appalachia | 3.89 |
| Southeast | 1.26 |
| Total | 47.58 |

Most of the HEL cropland was converted to CRP land between 1987 and 1997. The land use conversion drastically reduced the magnitude, severity, and rate of erosion. Drastic reduction of erosion on CRP-HEL improved soil C sequestration, decreased sediment load in rivers, and enhanced the quality of natural waters.

The soil conservation provisions of the 1985 act apply to all HEL. As per this act, any soil with an erodibility index of 8 or more has the potential to erode at least 8 times the rate at which it regenerates, and is thus highly erodible (Reichelderfer, 1987). There are at least 47.8 Mha of HEL, of which 14.1 Mha are being managed to prevent excessive erosion. The distribution of HEL area is shown in Table 6.13.

The CRP assists owners of HEL in conserving and improving soil and water resources of their farms and ranches. The HEL enrolled in CRP must be planted to grasses, shrubs, or trees and cannot be used for any commercial purpose, including grazing for at least 10 years. In return, the owners (operators) are compensated about 50% of the cost of establishing the grass or tree cover, and are also paid the annual rental payment. The rental payment ranges from $114/ha ($46/acre) to $222/ha ($90/acre).

There is also a land retirement program. The underlying premise of this program is that large public benefits can be gained by radically changing agricultural practices on a particular parcel of land. The payment mechanisms that can be used to facilitate land retirement strategies are lump sum payments or annual "rental fees" (Uri, 2001).

Both CRP and land retirement programs have been extremely successful in reducing risks of soil erosion from the privately owned land. These have been win–win strategies.

## REFERENCES

Hester, J.W., T.L. Thurow, and C.A. Taylor, Jr. Hydrologic characteristics of vegetation types as affected by prescribed burning. *J. Range Manage.* 50, 199–204, 1997.

Holechek, J.L., R.D. Pieper, and C.H. Herbel. *Range Management: Principles and Practices.* Regents/Prentice Hall. Englewood Cliffs, NJ, 1989.

Lal, R. Soil erosion and carbon dynamics on grazing land. In *The Potential of U.S. Grazing Lands to Sequester Carbon and Mitigate the Greenhouse Effect.* R.F. Follett, J.M. Kimble, and R. Lal (Eds.), CRC/Lewis Publishers, Boca Raton, FL, 2001a, pp. 231–247.

Lal, R. The physical quality of soil on grazing lands and its effects on sequestering carbon. In *The Potential of U.S. Grazing Lands to Sequester Carbon and Mitigate the Greenhouse Effect*. R.F. Follett, J.M. Kimble, and R. Lal (Eds.), CRC/Lewis Publishers, Boca Raton, FL, 2001b, pp. 249–265.

Reichelderfer, K. A farm program with incentives to do good. In *Our American Land, 1987 Yearbook of Agriculture*. USDA, Washington, D.C., 1987, pp. 267–271.

Sobecki, T.M., D.L. Moffitt, J. Stone, C.D. Franks, and A.G. Mendenhall. A broad-scale perspective on the extent, distribution, and characteristics of U.S. grazing land. In *The Potential of U.S. Grazing Lands to Sequester Carbon and Mitigate the Greenhouse Effect*. R.F. Follett, J.M. Kimble, and R. Lal (Eds.), CRC/Lewis Publishers, Boca Raton, FL, 2001, pp. 21–63.

Uri, N. A note on soil erosion and its environmental consequences in the United States. *Water, Air Soil Pollut.* 129, 181–197, 2001.

# CHAPTER 7

# Wind Erosion on Private Lands in the U.S.

In contrast to water erosion, which is caused by kinetic energy of flowing water and raindrop impact, wind erosion implies soil detachment and its transport by forces generated by wind (Plate 7.1 and Plate 7.2). The wind erosion hazard is severe in regions of low precipitation and high temperatures and wind velocity where soil is bare for an extended period. These conditions occur in arid and semiarid regions of the U.S. The risks of wind erosion are exacerbated by wind blowing across long, bare fields on soils of single-grain or weak structure and having a loamy texture.

Similar to water erosion, however, wind erosion is accentuated by anthropogenic perturbations. Activities that lead to soil disturbances and accentuate risks of wind erosion include clean cultivation, overgrazing, and fire. There are four factors affecting the severity of wind erosion: wind characteristics, meteorological factors, soil surface, and land use. The mechanics of wind erosion, its measurement, and its predictions are explained by Skidmore (1994) and Skidmore et al. (1994). Principles of wind erosion control are similar to those of water erosion and include establishment of ground cover, conversion of plow till to no till or conservation tillage, mulch farming, controlled grazing, wind breaks, grass strips, reduction of fallow, etc.

## 7.1 CROPLAND AREA AFFECTED BY WIND EROSION

The cropland area affected by wind erosion is shown in Table 7.1 for both cultivated and noncultivated categories. The land area affected by wind erosion (moderate +) on cultivated cropland was 31.5 Mha (20.7%) in 1982, 31.1 Mha (21.2%) in 1987, 21.4 Mha (15.7%) in 1992, and 19.2 Mha (14.5%) in 1997. The cultivated area affected by severe wind erosion decreased at an average rate of 0.8 Mha/yr or 2.6%/yr (Figure 7.1). Considering all severity classes of erosion (including light erosion), the land area affected by wind erosion on cultivated cropland was 72.1 Mha in 1982, 70.2 Mha in 1987, 60.8 Mha in 1992, and 57.9 Mha

**Plate 7.1**  Wind erosion occurs on plowed soil with lack of protective vegetal cover.

**Plate 7.2**  Severe case of wind erosion involves encroachment of vegetation by sand dune.

in 1997. Regression equations relating wind erosion hazard on cropland between 1982 and 1997 are shown in Equation 7.1 to Equation 7.3.

$$Y = 32.8 - 0.93x, \quad R^2 = 0.88 \qquad (7.1)$$

$$Y = 0.36 - 0.009x, \quad R^2 = 0.58 \qquad (7.2)$$

$$Y = 34.1 - 0.96x, \quad R^2 = 0.82 \qquad (7.3)$$

Table 7.1 Cropland Area Affected by Wind Erosion on Private U.S. Land

| Land Use | Year | Total Area | Light | Moderate | Severe | Extreme | Total (Moderate +) | % of Total Area |
|---|---|---|---|---|---|---|---|---|
| | | | | | Mha | | | |
| Cultivated | 1982 | 152.41 | 40.62 | 14.10 | 13.75 | 3.63 | 31.48 | 20.7 |
| | 1987 | 146.95 | 39.06 | 14.37 | 13.29 | 3.43 | 31.09 | 21.2 |
| | 1992 | 135.72 | 39.39 | 10.21 | 8.79 | 2.36 | 21.36 | 15.7 |
| | 1997 | 132.30 | 38.75 | 9.21 | 7.99 | 1.98 | 19.18 | 14.5 |
| Noncultivated | 1982 | 18.02 | 0.86 | 0.14 | 0.13 | 0.04 | 0.31 | 1.7 |
| | 1987 | 17.68 | 1.03 | 0.17 | 0.15 | 0.06 | 0.38 | 2.1 |
| | 1992 | 19.06 | 0.81 | 0.14 | 0.08 | 0.04 | 0.26 | 1.4 |
| | 1997 | 20.33 | 0.75 | 0.09 | 0.08 | 0.03 | 0.20 | 1.0 |

Total erosion area comprises sum of moderate, severe, and extreme forms, and % of the total is based on the total cropland area under that category.

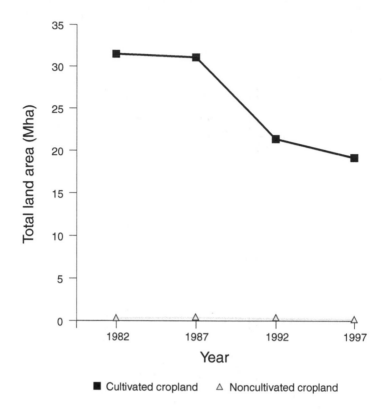

**Figure 7.1**   Land area affected by wind erosion (moderate and above) on U.S. cropland.

The area of cultivated cropland affected by wind erosion declined at a rate of 0.93 Mha/yr (Equation 7.1) for cultivated cropland, 0.009 Mha/yr (Equation 7.2) for noncultivated cropland, and 0.96 Mha/yr (Equation 7.3) for total cropland affected by erosion. Total land area affected by wind erosion at moderate and above severity of erosion is shown in Figure 7.2. The area affected declined from more than 32 Mha in 1982 to less than 20 Mha in 1997.

## 7.2  WIND EROSION ON IRRIGATED AND NONIRRIGATED CROPLAND

The nonirrigated cultivated cropland area was 128.8 Mha in 1982, 122.9 Mha in 1987, and 112.5 Mha in 1992. In contrast, the irrigated cultivated area for the same years was 19.5, 19.2, and 19.3 Mha, respectively. The noncultivated and nonirrigated cropland area was 16.7 Mha in 1982, 17.0 Mha in 1987, and 17.2 Mha in 1992. The noncultivated irrigated cropland area for the same period was 5.50, 5.60, and 5.9 Mha, respectively (Table 7.2). The cropland area affected by high erosion risks (moderate +) for nonirrigated cultivated land was 26.0 Mha (20.2%)

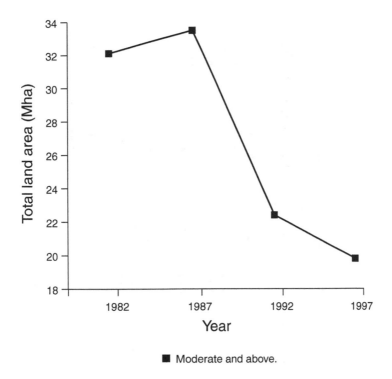

**Figure 7.2**  Total land area affected by wind erosion.

in 1982, 25.7 Mha (20.9%) in 1987, and 16.7 Mha (14.8%) in 1992. In comparison, the land area affected by all categories of wind erosion on cultivated nonirrigated cropland for the same years was 61.7, 59.5, and 50.9 Mha, respectively (Table 7.2).

Similar to the water erosion, the land area affected by wind erosion also declined over time. In comparison with nonirrigated cultivated land, the other cropland area affected by wind erosion was rather small. In these categories, the extent of land area affected by wind erosion also declined with time.

## 7.3 SEVERITY OF WIND EROSION ON U.S. CROPLAND

The data in Table 7.3 show erosion hazard on U.S. cropland in terms of the T value. The land area affected by wind erosion exceeding T value (>T) was rather small. Therefore, wind erosion on U.S. cropland is not a major issue. However, it was a major problem during the "Dust Bowl" era. It was the spectacular dust storm that led to the creation of the Soil Conservation Service (SCS) or the Natural Resources Conservation Service (NRCS). Air quality caused by wind-blown material remains a problem in several dry regions along the western edge of the Great Plains; however, the magnitude and severity of the problem is less.

Table 7.2  Irrigated and Nonirrigated Cropland Area Affected by Wind Erosion on U.S. Private Land (1992 Assessment)

| Land Use | Type | Year | Total Area | Area Affected by Wind Erosion | | | | | |
|---|---|---|---|---|---|---|---|---|---|
| | | | | Light | Moderate | Severe | Extreme | Total (Moderate+) | % of Total |
| | | | | Mha | | | | | |
| Cultivated | Nonirrigated | 1982 | 128.76 | 35.67 | 12.26 | 10.72 | 3.03 | 26.01 | 20.2 |
| | | 1987 | 122.87 | 33.81 | 12.26 | 10.60 | 2.82 | 25.68 | 20.9 |
| | | 1992 | 112.46 | 34.25 | 8.35 | 6.48 | 1.85 | 16.68 | 14.8 |
| | Irrigated | 1982 | 19.50 | 4.31 | 1.76 | 2.97 | 0.60 | 5.33 | 27.3 |
| | | 1987 | 19.20 | 4.62 | 1.88 | 2.75 | 0.60 | 5.23 | 27.2 |
| | | 1992 | 19.31 | 4.64 | 1.73 | 2.29 | 0.59 | 4.61 | 23.9 |
| Noncultivated | Nonirrigated | 1982 | 16.67 | 1.09 | 0.12 | 0.07 | 0.41 | 0.60 | 3.6 |
| | | 1987 | 16.97 | 1.26 | 0.19 | 0.08 | 0.79 | 1.06 | 6.2 |
| | | 1992 | 17.15 | 0.99 | 0.12 | 0.03 | 0.30 | 0.45 | 2.6 |
| | Irrigated | 1982 | 5.50 | 0.61 | 0.15 | 0.13 | 0.05 | 0.33 | 6.0 |
| | | 1987 | 5.60 | 0.67 | 0.12 | 0.16 | 0.07 | 0.35 | 6.3 |
| | | 1992 | 5.87 | 0.68 | 0.14 | 0.12 | 0.05 | 0.31 | 5.3 |

Total erosion area comprises sum of moderate, severe, and extreme forms, and % of the total is based on the total cropland area under that category.

Table 7.3 Severity of Annual Wind Erosion on Privately Owned U.S. Cropland

| Land Use | Year | ≤T | 1–2T | 2–3T | 3–4T | 4–5T | >5T | Total >T |
|----------|------|------|--------|-------|------|------|-------|-------|
| | | | | 10³ ha | | | | |
| Cultivated | 1982 | 298.79 | 34.74 | 17.89 | 8.96 | 4.74 | 11.35 | 77.68 |
| | 1987 | 286.27 | 35.40 | 16.94 | 8.93 | 4.97 | 10.45 | 76.69 |
| | 1992 | 282.59 | 25012.00 | 11.17 | 5.92 | 3.15 | 7.30 | 52.66 |
| | 1997 | 279.46 | 22.69 | 10.15 | 5.43 | 2.91 | 6.14 | 47.32 |
| Noncultivated | 1982 | 43.74 | 0.34 | 0.015 | 0.07 | 0.05 | 0.15 | 0.76 |
| | 1987 | 42.74 | 0.41 | 0.21 | 0.11 | 0.06 | 0.15 | 0.93 |
| | 1992 | 46.42 | 0.35 | 0.11 | 0.04 | 0.04 | 0.10 | 0.66 |
| | 1997 | 49.71 | 0.23 | 0.12 | 0.04 | 0.03 | 0.08 | 0.50 |

## 7.4 WIND EROSION ON U.S. CROPLAND FOR DIFFERENT SEVERITY CLASSES

The data in Table 7.4 show wind erosion hazard on cropland for different severity classes (e.g., slight, moderate, and severe) for all states in the conterminous U.S. Cropland highly susceptible to wind erosion occurs in Texas (1.6 Mha), Colorado (0.54 Mha), Montana (0.3 Mha), Minnesota (0.25 Mha), Wyoming (0.18 Mha), New Mexico (0.17 Mha), and Washington (0.12 Mha). Cropland area in these states needs to be protected by conversion to a restorative land use and adoption of conservation effective farming systems.

## 7.5 WIND EROSION ON GRAZING LAND

Grazing land comprises pastureland, rangeland, and forestland. The extent of wind erosion on pastureland is shown by the data in Table 7.5. The land area affected by moderate and above categories of wind erosion on pastureland ranged between 0.4 and 0.5% of the total area, and this area decreased over the 15-year period.

The data in Table 7.6 show wind erosion hazard on grazed and nongrazed pastureland, rangeland, and forestland. Similar to water erosion, there was also no wind erosion on grazed or nongrazed forestlands reported, though the potential exists but at a very low level. Regardless of grazing, the extent of area affected by accelerated wind erosion (moderate +) ranged between 0.4 and 0.5% of the total area.

As one would expect, wind erosion is most prevalent on rangeland. The land area affected by accelerated wind erosion (moderate +) on grazed and nongrazed rangeland varied between 13.7 and 15.9%. Severity of wind erosion on U.S. grazing land, both pastureland and rangeland, is shown in Table 7.7. The data showed that wind erosion may be a hazard in isolated places but is not a serious issue at national level in pastureland or grazing land.

Table 7.4  U.S. Cropland Area Subjected to Different Severity of Wind Erosion

| State | Slightly Eroded (<T) | Moderately Eroded (1–4T) | Severely Eroded (>4T) |
|---|---|---|---|
| | 1000 ha | | |
| Alabama | 1274.0 | 0 | 0 |
| Arizona | 267.4 | 133.0 | 84.4 |
| Arkansas | 3129.5 | 0 | 0 |
| California | 3984.7 | 56.6 | 28.3 |
| Colorado | 1618.6 | 1458.8 | 541.9 |
| Connecticut | 92.5 | 0 | 0 |
| Delaware | 198.5 | 3.5 | 0 |
| Florida | 1212.6 | 0.9 | — |
| Georgia | 2094.3 | 0 | 0 |
| Idaho | 1591.7 | 617.0 | 58.5 |
| Illinois | 9757.0 | 0 | 0 |
| Indiana | 5372.8 | 97.8 | 0 |
| Iowa | 9455.5 | 658.4 | 2.6 |
| Kansas | 9560.2 | 1121.9 | 72.7 |
| Kentucky | 2061.5 | 0 | 0 |
| Louisiana | 2417.7 | 0 | 0 |
| Maine | 181.2 | 0 | 0 |
| Maryland | 677.4 | 0 | 0 |
| Massachusetts | 110.2 | 0 | 0 |
| Michigan | 3228.6 | 396.6 | 12.5 |
| Minnesota | 5208.8 | 3182.6 | 254.7 |
| Missouri | 1334.6 | 0 | 0 |
| Montana | 3601.5 | 2184.7 | 300.7 |
| Nebraska | 7165.9 | 583.8 | 39.5 |
| Nevada | 249.2 | 29.6 | 29.8 |
| New Hampshire | 57.3 | 0 | 0 |
| New Jersey | 263.0 | 0 | 0 |
| New Mexico | 330.9 | 262.8 | 172.1 |
| New York | 2273.7 | 0 | 0 |
| North Carolina | 2412.8 | 0 | 0 |
| North Dakota | 8909.9 | 1023.7 | 83.8 |
| Ohio | 4823.8 | 4.8 | 0.8 |
| Oklahoma | 3668.7 | 377.8 | 34.8 |
| Oregon | 1468.9 | 44.3 | 15.3 |
| Pennsylvania | 2265.5 | 0 | 0 |
| Rhode Island | 10.1 | 0 | 0 |
| South Carolina | 1207.5 | 0 | 0 |
| South Dakota | 5713.2 | 920.6 | 20.5 |
| Tennessee | 1966.3 | 0 | 0 |
| Texas | 7004.3 | 2791.0 | 1646.6 |
| Utah | 580.0 | 110.8 | 44.0 |
| Vermont | 256.9 | 0 | 0 |
| Virginia | 1172.8 | 1.0 | 0.7 |
| Washington | 2121.9 | 486.8 | 122.1 |
| West Virginia | 370.3 | 0 | 0 |
| Wisconsin | 4326.2 | 50.5 | 1.1 |
| Wyoming | 566.9 | 172.1 | 180.6 |
| Total | 134,004.0 (±691.5) | 16,771.6 (±344.4) | 3748.0 (±168.2) |

Source: From NRI, National Resources Inventory, NRCS, Washington, D.C., 1992.

**Table 7.5 Land Area under Pasture Affected by Wind Erosion on Private U.S. Land**

| | | Area Affected by Wind Erosion | | | | | |
|---|---|---|---|---|---|---|---|
| Year | Total Area | Light | Moderate | Severe | Extreme | Total Area (Moderate+) | % of Total Area |
| | | | | Mha | | | |
| 1982 | 53.44 | 0.64 | 0.12 | 0.098 | 0.055 | 0.27 | 0.5 |
| 1987 | 51.87 | 0.70 | 0.11 | 0.094 | 0.061 | 0.27 | 0.5 |
| 1992 | 51.03 | 0.59 | 0.12 | 0.071 | 0.044 | 0.23 | 0.5 |
| 1997 | 48.58 | 0.64 | 0.09 | 0.068 | 0.031 | 0.19 | 0.4 |

## 7.6 WIND EROSION ON CRP LAND

The area prone to wind erosion (moderate+) on CRP land was 1.81 Mha (32.4%) in 1987, 0.52 Mha (3.8%) in 1992, and 0.17 Mha (1.3%) in 1997 (Table 7.8). The wind erosion hazard decreased with time even on CRP land. The decrease in wind erosion hazard on CRP land between 1987 and 1997 was due to establishment of vegetative cover that was effective in minimizing the wind erosion risks.

## 7.7 WIND EROSION ON MINOR MISCELLANEOUS LAND

The area under minor land use ranged between 21.4 Mha and 22.1 Mha, and increased over the 10-year period between 1982 and 1992. The area affected by accelerated erosion (>T) for other lands ranged between 1.2 Mha and 1.3 Mha or about 6% of the total area (Table 7.9).

## 7.8 COMPARISON OF WIND EROSION HAZARD AMONG PREDOMINANT LAND USES

The data in Table 7.10 show the magnitude of wind erosion hazard among different land uses in each state within the conterminous U.S. Of the total wind erosion estimated at 2.6 billion Mg (1992 NRI data), 832 million Mg was from cultivated cropland, 115 million Mg from noncultivated cropland, 9.1 billion million Mg from pastureland, 1.6 billion Mg from rangeland, and 163 million Mg from other land uses. In rangeland, the most severe wind erosion is observed in New Mexico, Arizona, Wyoming, Colorado, and Utah. In cropland, most severe wind erosion is observed in Texas, Minnesota, Colorado, and Montana (Table 7.10).

## 7.9 TOTAL LAND AREA ON U.S. PRIVATE LAND AFFECTED BY WIND EROSION

The data in Table 7.11 and Table 7.12 show the total land area affected by wind erosion (moderate +) was 32.1 Mha (5.6%) in 1982, 33.5 Mha (5.9%) in 1987, 22.4 Mha (3.9%) in 1992, and 19.8 Mha (3.5%) in 1997 (Table 7.11). The land area

Table 7.6   Grazing Land Area Affected by Wind Erosion on U.S. Private Land (1992 Assessment)

| Land Use | Type | Year | Total Area | Area Affected by Wind Erosion | | | | | |
|---|---|---|---|---|---|---|---|---|---|
| | | | | Light | Moderate | Severe | Extreme | Total (Moderate+) | % of Total |
| | | | | | | Mha | | | |
| Pastureland | Grazed | 1982 | 40.16 | 0.43 | 0.089 | 0.076 | 0.037 | 0.20 | 0.5 |
| | | 1987 | 42.73 | 0.52 | 0.097 | 0.082 | 0.049 | 0.23 | 0.5 |
| | | 1992 | 45.78 | 0.19 | 0.122 | 0.067 | 0.039 | 0.28 | 0.6 |
| | Nongrazed | 1982 | 13.23 | 0.17 | 0.045 | 0.033 | 0.022 | 0.10 | 0.7 |
| | | 1987 | 8.93 | 0.14 | 0.025 | 0.022 | 0.015 | 0.062 | 0.7 |
| | | 1992 | 5.20 | 0.07 | 0.007 | 0.009 | 0.006 | 0.022 | 0.4 |
| Rangeland | Grazed | 1982 | 155.04 | 15.09 | 5.39 | 9.77 | 9.52 | 24.62 | 15.9 |
| | | 1987 | 155.43 | 16.56 | 5.54 | 9.51 | 9.64 | 24.69 | 15.9 |
| | | 1992 | 156.36 | 16.70 | 5.67 | 9.25 | 9.59 | 24.51 | 15.7 |
| | Nongrazed | 1982 | 10.51 | 0.89 | 0.32 | 0.86 | 0.44 | 1.62 | 15.4 |
| | | 1987 | 7.65 | 0.49 | 0.18 | 0.58 | 0.29 | 1.05 | 13.7 |
| | | 1992 | 5.16 | 0.29 | 0.08 | 0.43 | 0.21 | 0.72 | 14.0 |
| Forestland | Grazed | 1982 | 26.59 | 0 | 0 | 0 | 0 | 0 | 0 |
| | | 1987 | 25.72 | 0 | 0 | 0 | 0 | 0 | 0 |
| | | 1992 | 25.49 | 0 | 0 | 0 | 0 | 0 | 0 |
| | Nongrazed | 1982 | 133.07 | 0 | 0 | 0 | 0 | 0 | 0 |
| | | 1987 | 134.54 | 0 | 0 | 0 | 0 | 0 | 0 |
| | | 1992 | 133.07 | 0 | 0 | 0 | 0 | 0 | 0 |

Total erosion area comprises sum of moderate, severe, and extreme forms, and % of the total is based on the total cropland area under that category.

Table 7.7   Severity of Annual Wind Erosion on Privately Owned U.S. Grazing Land

| Land Use | Year | ≤T | 1–2T | 2–3T | 3–4T | 4–5T | >5T | >T Total Area |
|----------|------|-----|------|------|------|------|-----|---------------|
| | | | | | 10³ ha | | | |
| Pastureland | 1982 | 131.33 | 0.29 | 0.15 | 0.05 | 0.03 | 0.14 | 0.70 |
| | 1987 | 127.45 | 0.28 | 0.15 | 0.05 | 0.06 | 0.13 | 0.7 |
| | 1992 | 125.48 | 0.29 | 0.09 | 0.07 | 0.03 | 0.11 | 0.6 |
| | 1997 | 119.52 | 0.22 | 0.10 | 0.04 | 0.03 | 0.08 | 0.5 |
| Rangeland | 1982 | 416.74 | 0 | 0 | 0 | 0 | 0 | 0 |
| | 1987 | 405.26 | 0 | 0 | 0 | 0 | 0 | 0 |
| | 1992 | 405.21 | 0 | 0 | 0 | 0 | 0 | 0 |
| | 1997 | 405.98 | 0 | 0 | 0 | 0 | 0 | 0 |

TABLE 7.8   Land Area under CRP Affected by Wind Erosion on Private U.S. Land

| | | Area Affected by Wind Erosion | | | | | |
|------|-------|-------|----------|--------|---------|------------------|------------|
| Year | Total Area | Light | Moderate | Severe | Extreme | Total (Moderate +) | % of Total Area |
| | | | | Mha | | | |
| 1987 | 5.59 | 1.13 | 0.54 | 0.97 | 0.30 | 1.81 | 32.4 |
| 1992 | 13.78 | 1.36 | 0.30 | 0.15 | 0.07 | 0.52 | 3.8 |
| 1997 | 13.24 | 0.86 | 0.09 | 0.05 | 0.03 | 0.17 | 1.3 |

Total erosion area comprises sum of moderate, severe, and extreme forms, and % of the total is based on the total cropland area under that category.

Table 7.9   Land Area Affected by Wind Erosion on Minor Land on Private U.S. Land (1992 Assessment)

| | | Area Affected by Wind Erosion | | | | | |
|------|-------|-------|----------|--------|---------|------------------|------------|
| Year | Total Area | Light | Moderate | Severe | Extreme | Total (Moderate +) | % of Total Area |
| | | | | Mha | | | |
| 1982 | 21.39 | 0.34 | 0.15 | 0.77 | 0.30 | 1.22 | 5.7 |
| 1987 | 21.58 | 0.39 | 0.18 | 0.76 | 0.31 | 1.25 | 5.8 |
| 1992 | 22.10 | 0.36 | 0.18 | 0.83 | 0.31 | 1.32 | 6.0 |

Total erosion area comprises sum of moderate, severe, and extreme forms, and % of the total is based on the total cropland area under that category.

affected by the wind erosion hazard decreased over the 15-year period, at an average rate of 0.8 Mha/yr. The area affected by extreme category of wind erosion was only 3.7 Mha in 1982, 3.9 Mha in 1987, 2.5 Mha in 1992, and 2.1 Mha in 1997.

The wind erosion hazard based on the 1992 assessment is different than that on the 1997 assessment. Wind erosion hazard declined in 1992 due to adoption of judicious land use and best management practices (BMPs) on cropland and grazing land. Distribution of the wind erosion hazard on all privately owned U.S. land in terms of T values shown in Table 7.12 reveals that less than 2 Mha of area is affected by wind erosion.

Table 7.10 The Magnitude of Wind Erosion in Relation to Land Use for States in the Conterminous U.S.

| State | Cropland | | Forestland | Pastureland | Rangeland | Other | Total |
|---|---|---|---|---|---|---|---|
| | Cultivated | Noncultivated | | | | | |
| | | | | 1000 Mg/yr | | | |
| Alabama | 0.0 | 0.0 | 0.0 | 0.0 | 0.0 | 0.0 | 0.0 |
| Arizona | 14550.3 | 43.0 | 0.0 | 20.0 | 283217.1 | 12086.5 | 309917.0 |
| Arkansas | 0.0 | 0.0 | 0.0 | 0.0 | 0.0 | 0.0 | 0.0 |
| California | 5655.8 | 900.5 | 0.0 | 325.4 | 47091.5 | 32473.6 | 86446.8 |
| Colorado | 80106.8 | 1126.7 | 0.0 | 2726.9 | 174586.3 | 11446.8 | 269993.4 |
| Connecticut | 0.0 | 0.0 | 0.0 | 0.0 | 0.0 | 0.0 | 0.0 |
| Delaware | 678.9 | 0.0 | 0.0 | 0.2 | 0.0 | 84.4 | 763.5 |
| Florida | 0.0 | 33.1 | 0.0 | 2.8 | 0.0 | 0.0 | 35.9 |
| Georgia | 0.0 | 0.0 | 0.0 | 0.0 | 0.0 | 0.0 | 0.0 |
| Idaho | 21667.8 | 234.3 | 0.0 | 186.4 | 19578.0 | 1803.2 | 43469.8 |
| Illinois | 0.0 | 0.0 | 0.0 | 0.0 | 0.0 | 0.0 | 0.0 |
| Indiana | 4471.1 | 6.2 | 0.0 | 0.0 | 0.0 | 67.9 | 4545.1 |
| Iowa | 30639.5 | 52.5 | 0.0 | 31.1 | 0.0 | 366.2 | 31089.3 |
| Kansas | 50484.6 | 922.4 | 0.0 | 54.4 | 9599.6 | 3199.3 | 64260.5 |
| Kentucky | 0.0 | 0.0 | 0.0 | 0.0 | 0.0 | 0.0 | 0.0 |
| Louisiana | 0.0 | 0.0 | 0.0 | 0.0 | 0.0 | 0.0 | 0.0 |
| Maine | 0.0 | 0.0 | 0.0 | 0.0 | 0.0 | 0.0 | 0.0 |
| Maryland | 107.4 | 0.0 | 0.0 | 0.0 | 0.0 | 1.5 | 108.9 |
| Massachusetts | 0.0 | 0.0 | 0.0 | 0.0 | 0.0 | 0.0 | 0.0 |
| Michigan | 16687.1 | 420.7 | 0.0 | 131.9 | 0.0 | 293.7 | 17533.5 |
| Minnesota | 109439.1 | 331.3 | 0.0 | 412.1 | 0.0 | 314.7 | 110497.1 |
| Mississippi | 0.0 | 0.0 | 0.0 | 0.0 | 0.0 | 0.0 | 0.0 |
| Missouri | 0.0 | 0.0 | 0.0 | 0.0 | 0.0 | 0.0 | 0.0 |
| Montana | 83945.8 | 296.1 | 0.0 | 243.2 | 17255.6 | 2505.5 | 104246.2 |
| Nebraska | 27176.7 | 127.4 | 0.0 | 147.6 | 3160.0 | 2447.7 | 33059.5 |
| Nevada | 8824.7 | 1072.2 | 0.0 | 406.0 | 67127.3 | 6481.9 | 83912.1 |
| New Hampshire | 0.0 | 0.0 | 0.0 | 0.0 | 0.0 | 0.0 | 0.0 |
| New Jersey | 30.1 | 5.4 | 0.0 | 0.1 | 0.0 | 80.9 | 116.5 |

| | | | | | | | |
|---|---|---|---|---|---|---|---|
| New Mexico | 22066.5 | 1186.0 | 0.0 | 844.1 | 460929.8 | 14641.8 | 499668.1 |
| New York | 1.9 | 0.0 | 0.0 | 0.0 | 0.0 | 0.0 | 1.9 |
| North Carolina | 7.3 | 0.0 | 0.0 | 0.0 | 0.0 | 0.0 | 7.3 |
| North Dakota | 38754.0 | 102.2 | 0.0 | 2.6 | 215.9 | 896.2 | 39971.0 |
| Ohio | 990.0 | 2.0 | 0.0 | 0.0 | 0.0 | 0.0 | 992.0 |
| Oklahoma | 15559.3 | 85.8 | 0.0 | 76.6 | 5251.5 | 1538.1 | 22511.3 |
| Oregon | 3975.5 | 65.8 | 0.0 | 177.9 | 2226.2 | 917.6 | 7363.0 |
| Pennsylvania | 0.0 | 0.0 | 0.0 | 0.0 | 0.0 | 0.0 | 0.0 |
| Rhode Island | 0.0 | 0.0 | 0.0 | 0.0 | 0.0 | 0.0 | 0.0 |
| South Carolina | 0.0 | 0.0 | 0.0 | 0.0 | 0.0 | 0.0 | 0.0 |
| South Dakota | 34474.2 | 514.9 | 0.0 | 199.8 | 1458.3 | 405.4 | 37052.7 |
| Tennessee | 0.0 | 0.0 | 0.0 | 0.0 | 0.0 | 0.0 | 0.0 |
| Texas | 208216.1 | 957.0 | 0.0 | 403.5 | 96533.2 | 36185.6 | 342295.5 |
| Utah | 5805.0 | 1012.3 | 0.0 | 908.5 | 115257.7 | 29892.1 | 152875.6 |
| Vermont | 0.0 | 0.0 | 0.0 | 0.0 | 0.0 | 0.0 | 0.0 |
| Virginia | 405.1 | 0.1 | 0.0 | 3.9 | 0.0 | 12.8 | 421.8 |
| Washington | 27654.7 | 600.0 | 0.0 | 169.3 | 10593.0 | 684.4 | 39701.5 |
| West Virginia | 0.0 | 0.0 | 0.0 | 0.0 | 0.0 | 0.0 | 0.0 |
| Wisconsin | 1817.1 | 39.1 | 0.0 | 32.2 | 0.0 | 18.1 | 1906.5 |
| Wyoming | 18141.2 | 1396.0 | 0.0 | 1596.1 | 274374.7 | 4633.8 | 300141.9 |
| Total | 832334.0 | 11532.9 | 0.0 | 9102.8 | 1588455.5 | 163479.7 | 2604905.0 |

Source: From NRI, National Resources Inventory, NRCS, Washington, D.C., 1992.

Table 7.11   Total Land Area Affected by Wind Erosion on Private U.S. Land

| Year | Total Area | Area Affected by Wind Erosion | | | | Total (Moderate +) | % of Total Area |
|------|-----------|-------|----------|--------|---------|------------------|-----------------|
|      |           | Light | Moderate | Severe | Extreme |                  |                 |
|      |           |       |          | Mha    |         |                  |                 |
| 1982 | 575.99 | 42.11 | 14.36 | 13.98 | 3.72 | 32.06 | 5.6 |
| 1987 | 572.81 | 41.93 | 15.18 | 14.50 | 3.86 | 33.54 | 5.9 |
| 1992 | 579.08 | 42.15 | 10.77 | 9.09 | 2.51 | 22.37 | 3.9 |
| 1997 | 564.28 | 41.00 | 9.49 | 8.19 | 2.07 | 19.75 | 3.5 |

Total erosion area comprises sum of moderate, severe and extreme forms, and % of the total is based on the total cropland area under that category.

*Source*: From NRI, National Resources Inventory, NRCS, Washington, D.C., 1997.

Table 7.12   Severity of Annual Wind Erosion on All Privately Owned U.S. Land

| Year | ≤T | 1–2T | 2–3T | 3–4T | 4–5T | >5T |
|------|------|------|------|------|------|------|
|      |      |      | 10³ ha |      |      |      |
| 1982 | 1865 | 35 | 18 | 9.1 | 4.8 | 11.6 |
| 1987 | 1861 | 37 | 18 | 9.6 | 5.6 | 12.0 |
| 1992 | 1889 | 26 | 12 | 6.1 | 3.3 | 7.7 |
| 1997 | 1895 | 23 | 10 | 5.5 | 3.0 | 6.4 |

## 7.10 CONCLUSIONS

Similar to water erosion, wind erosion hazard is decreasing on all land use categories. Conversion of HEL lands to CRP and other restorative land uses has enhanced the vegetal cover and decreased wind erosion risks. Conversion of plow till to conservation till, creation of wind breaks, improvements in pastureland, and reduction of summer fallow have decreased soil erosion risks.

## REFERENCES

NRI. National Resources Inventory, NRCS, Washington, D.C., 1997.

Skidmore, E.L. Wind erosion. In *Soil Erosion Research Methods*. R. Lal (Ed.), Soil Water Cons. Soc., Ankeny, IA, 1994, 265–293.

Skidmore, E.L., L.J. Hagen, D.V. Armbust, A.A. Durar, D.W. Fryrear, K.N. Potter, L.E. Wagner, and T.M. Zobeck. Methods of investigating basic processes and conditions affecting wind erosion. In *Soil Erosion Research Methods*. R. Lal (Ed.), Soil Water Cons. Soc., Ankeny, IA, 1994, 295–330.

# CHAPTER 8

# Water and Wind Erosion on U.S. Cropland

Both water and wind erosion can be serious problems on cropland, especially on land that is managed by intensive tillage methods of seedbed preparation. The most predominant form of water erosion is sheet or rill erosion, which occurs on soils even with gentle slopes. However, the extent (land area affected), magnitude (amount of soil displaced), and the rate (amount of soil displaced per unit area and per unit time) of sheet erosion are vastly accentuated by anthropogenic perturbations. In addition, rill erosion becomes a problem on steeplands with concentrated overland flow. Rills are shallow channels of concentrated flow and can easily be covered by plowing, but without proper management can easily become gullies. Gully erosion, a most spectacular form of water erosion, causes terrain deformation. Gullies are deep, often permanent, and not covered by plowing and other farm operations. In addition to water erosion, the displacem1ent of soil material by wind is often a problem in arid and semiarid regions. The extent, magnitude, and rate of wind erosion as well as water erosion are also accentuated by anthropogenic activities such as overgrazing, biomass burning, tillage, and other activities that lead to removal of the protective vegetation cover. Coarse-textured soils (e.g., sandy, sandy loam, loam sand, etc.) are more susceptible to wind erosion than fine-textured soils. Excessive wind erosion can also lead to terrain deformation.

Soils in some regions are susceptible both to water and wind erosion. Some soils in semiarid regions are prone to water erosion during the rainy season and wind erosion during the dry season.

The combined area, affected by both water and wind erosion, is of interest because of the off-site damages with negative impact on the environment. In addition, erosion also leads to decline in productivity (Lal, 1998; den Biggelaar et al., 2001). Impairment of water resources is the most widespread off-site impact of erosion (NRC, 1993; NRCS, 1995). The principal causes of pollution/contamination of surface waters include sedimentation, siltation, and eutrophication of rivers, waterways, lakes, and other natural water bodies (USEPA, 1995). Similar to water erosion, wind erosion also causes off-site damages (Piper and Lee, 1989). Important among these are air pollution, deposition of silt on urban centers, and contamination of water.

While the on-site impact leads to a temporary or permanent decline in agronomic/ biomass productivity, the off-site impact of water and wind erosion are more severe (Foster and Dabney, 1995; Crosson and Anderson, 2000). Therefore, the information about the extent and severity of land area affected by both water and wind erosion is of importance to society. Because of the severe off-site effects, society has a larger incentive to control erosion than farmers have (Crosson and Anderson, 2000).

## 8.1 THE EXTENT OF SHEET, RILL, AND WIND EROSION ON CROPLAND

### 8.1.1 Total Cropland

Total cropland area can be divided into highly erodible land (HEL) and nonhighly erodible land (NHEL) (Table 8.1), cultivated (Table 8.2), noncultivated (Table 8.3), and prime vs. nonprime land (Table 8.4). The extent of area affected by water and wind erosion on total cropland is shown in Table 8.1. The land area affected by severe erosion (>T) was 68.4 Mha (40.2%) in 1982, 64.1 Mha (38.9%) in 1987, 42.7 Mha (27.6%) in 1992, and 43.7 Mha (28.6%) in 1997.

### 8.1.2 Total Cultivated Land

The data in Table 8.2 show the extent of cultivated land area affected by sheet, rill, and wind erosion. The total land area affected was 67.5 Mha (44.3%) in 1982, 63.0 Mha (42.9%) in 1987, 47.4 Mha (34.9%) in 1992, and 42.6 Mha (32.2%) in 1997 (Figure 8.1). The adoption of best management practices (BMPs) and conversion to conservation-effective land use decreased the extent of total erosion (>T) on cultivated land over the 15-year period. The mean rate of decline was 1.7 Mha/yr. The decline in land area affected by erosion in different severity classes occurred in the order 1–2T > 2–3T > 3–4T > 5T > 4–5T.

Cultivated cropland area was divided into HEL and NHEL areas. Similar to the trends in total cultivated area, the extent of severe erosion also decreases in HEL and NHEL land over the 15-year period. The average rate of decline in land area affected by erosion was 0.7 Mha/yr or 2.2%/yr for HEL, and 0.9 Mha/yr or 2.7%/yr for NHEL (Figure 8.2; Table 8.2).

The data in Table 8.3 show the extent of land area affected by wind and water erosion on noncultivated land. In contrast to the cultivated land, there was no definite trend in the extent of erosion on noncultivated land over the 15-year period. The total land area affected by erosion ranged between 0.9 and 1.1 Mha, comprising 0.74 to 0.95 Mha on HEL, and 0.13 to 0.18 Mha on NHEL category.

### 8.1.3 Prime Land

Cropland can also be divided into prime land (with few or no limitations) and nonprime land (with some limitations) to cultivation for crop production. The extent of erosion on both prime and nonprime cropland decreased over the 15-year period.

Table 8.1 Total Soil Loss by Water and Wind Erosion on U.S. Cropland

| Land Use | Year | Total Area | ≤T | 1–2T | 2–3T | 3–4T | 4–5T | >5T | Total Area >T | >T as % of Total Area |
|---|---|---|---|---|---|---|---|---|---|---|
| | | | | | Mha | | | | | |
| Total cropland | 1982 | 170.4 | 101.9 | 32.6 | 14.3 | 7.2 | 4.1 | 10.2 | 68.4 | 40.2 |
| | 1987 | 164.6 | 100.5 | 31.2 | 13.2 | 7.0 | 4.0 | 8.7 | 64.1 | 38.9 |
| | 1992 | 154.8 | 106.5 | 25.3 | 9.9 | 4.9 | 2.6 | 5.7 | 48.4 | 27.6 |
| | 1997 | 152.6 | 109.0 | 24.0 | 8.8 | 4.2 | 2.2 | 4.5 | 43.7 | 28.6 |
| HEL cropland | 1982 | 50.4 | 16.5 | 8.6 | 6.8 | 5.0 | 3.5 | 9.9 | 33.8 | 67.1 |
| | 1987 | 47.3 | 15.3 | 8.7 | 6.8 | 4.8 | 3.2 | 8.4 | 31.9 | 67.4 |
| | 1992 | 42.6 | 16.8 | 8.4 | 6.0 | 3.7 | 2.2 | 5.6 | 25.9 | 60.8 |
| | 1997 | 42.1 | 18.9 | 8.8 | 5.3 | 3.1 | 1.7 | 4.3 | 23.2 | 55.1 |
| NHEL cropland | 1982 | 120.0 | 85.4 | 23.9 | 7.5 | 2.2 | 0.6 | 0.3 | 34.5 | 28.8 |
| | 1987 | 117.3 | 85.2 | 22.6 | 6.4 | 2.1 | 0.8 | 0.3 | 32.2 | 27.5 |
| | 1992 | 112.2 | 89.7 | 16.8 | 3.9 | 1.2 | 0.4 | 0.1 | 22.4 | 20.0 |
| | 1997 | 110.5 | 90.1 | 15.1 | 3.5 | 1.2 | 0.4 | 0.2 | 20.5 | 18.6 |

**Table 8.2  The Extent of Sheet, Rill, and Wind Erosion on Total Cultivated Land in Terms of Severity Classes**

| Land Use | Year | Total Area Affected | ≤T | 1–2T | 2–3T | 3–4T | 4–5T | >5T | Total Area >T | >T as % of Total Area |
|---|---|---|---|---|---|---|---|---|---|---|
| | | | | | Mha | | | | | |
| Total cultivated | 1982 | 152.34 | 84.87 | 32.07 | 14.14 | 7.11 | 4.08 | 10.09 | 67.48 | 44.3 |
| | 1987 | 146.89 | 83.94 | 30.63 | 13.02 | 6.84 | 3.91 | 8.55 | 62.95 | 42.9 |
| | 1992 | 135.67 | 88.3 | 24.72 | 9.71 | 4.78 | 2.54 | 5.61 | 47.36 | 34.9 |
| | 1997 | 132.24 | 89.61 | 23.4 | 8.6 | 4.17 | 2.13 | 4.34 | 42.64 | 32.2 |
| HEL cultivated | 1982 | 43.37 | 10.32 | 8.21 | 6.68 | 4.92 | 3.44 | 9.81 | 33.05 | 76.2 |
| | 1987 | 40.47 | 9.48 | 8.22 | 6.62 | 4.72 | 3.16 | 8.28 | 30.99 | 76.6 |
| | 1992 | 35.01 | 9.97 | 8.04 | 5.78 | 3.6 | 2.13 | 5.49 | 25.04 | 71.5 |
| | 1997 | 33.81 | 11.47 | 8.37 | 5.1 | 3.02 | 1.7 | 4.15 | 22.34 | 66.1 |
| NHEL cultivated | 1982 | 108.97 | 74.54 | 23.86 | 7.46 | 2.19 | 0.64 | 0.28 | 34.43 | 31.6 |
| | 1987 | 106.41 | 74.46 | 22.42 | 6.4 | 2.11 | 0.75 | 0.27 | 31.95 | 30.0 |
| | 1992 | 100.66 | 78.33 | 16.68 | 3.93 | 1.18 | 0.41 | 0.12 | 22.33 | 22.2 |
| | 1997 | 98.44 | 78.14 | 15.03 | 3.49 | 1.15 | 0.43 | 0.19 | 20.3 | 20.6 |

WATER AND WIND EROSION ON U.S. CROPLAND 113

Table 8.3  The Extent of Sheet, Rill, and Wind Erosion on Noncultivated Land in Terms of Severity Classes

| Land Use | Year | Total Area | ≤T | 1–2T | 2–3T | 3–4T | 4–5T | >5T | Total Area >T | >T as % of Total Area |
|---|---|---|---|---|---|---|---|---|---|---|
| | | Mha | | | | 10³ ha | | | | |
| Total noncultivated | 1982 | 18.01 | 17.02 | 520 | 200 | 90 | 50 | 130 | 990 | 5.5 |
| | 1987 | 17.67 | 16.55 | 570 | 230 | 120 | 60 | 150 | 1130 | 6.4 |
| | 1992 | 19.05 | 18.13 | 510 | 190 | 80 | 40 | 100 | 920 | 4.8 |
| | 1997 | 20.32 | 19.35 | 550 | 200 | 70 | 40 | 110 | 970 | 4.8 |
| HEL noncultivated | 1982 | 7.05 | 6.2 | 430 | 170 | 80 | 40 | 130 | 850 | 12.0 |
| | 1987 | 6.81 | 5.86 | 440 | 200 | 100 | 50 | 150 | 950 | 13.9 |
| | 1992 | 7.53 | 6.78 | 370 | 170 | 70 | 30 | 100 | 740 | 9.9 |
| | 1997 | 8.31 | 7.47 | 450 | 180 | 70 | 40 | 110 | 840 | 10.1 |
| NHEL noncultivated | 1982 | 10.96 | 10.82 | 10 | 30 | 10 | 0 | 0 | 140 | 1.3 |
| | 1987 | 10.87 | 10.69 | 130 | 30 | 10 | 0 | 0 | 180 | 1.7 |
| | 1992 | 11.52 | 11.35 | 140 | 20 | 10 | 10 | 0 | 180 | 1.5 |
| | 1997 | 12.01 | 11.88 | 100 | 20 | 0 | 0 | 0 | 130 | 1.1 |

Table 8.4 The Extent of Sheet, Rill, and Wind Erosion on Prime Agricultural Land in Relation to Severity Classes

| Land Use | Year | Total Area | ≤T | 1–2T | 2–3T | 3–4T | 4–5T | >5T | Total Area >T | >T as % of Total Area |
|---|---|---|---|---|---|---|---|---|---|---|
| | | | | | Mha | | | | | |
| Total prime cropland | 1982 | 93.43 | 62.19 | 18.67 | 6.98 | 2.8 | 1.28 | 1.51 | 31.24 | 33.4 |
| | 1987 | 91.31 | 62.74 | 17.47 | 6.04 | 2.63 | 1.22 | 1.21 | 28.57 | 31.3 |
| | 1992 | 87.41 | 66.21 | 13.57 | 4.39 | 1.75 | 0.7 | 0.8 | 21.21 | 24.3 |
| | 1997 | 85.91 | 66.86 | 12.44 | 3.76 | 1.55 | 0.66 | 0.64 | 19.04 | 22.2 |
| Prime cultivated | 1982 | 86.81 | 55.75 | 18.56 | 6.95 | 2.79 | 1.27 | 1.5 | 31.06 | 35.8 |
| | 1987 | 84.81 | 56.43 | 17.35 | 6.0 | 2.61 | 1.22 | 1.2 | 28.38 | 33.5 |
| | 1992 | 80.38 | 59.35 | 13.45 | 4.36 | 1.73 | 0.7 | 0.79 | 21.03 | 26.2 |
| | 1997 | 78.44 | 59.56 | 12.33 | 3.73 | 1.54 | 0.65 | 0.63 | 18.89 | 24.1 |
| Prime noncultivated | 1982 | 6.61 | 6.44 | 0.11 | 0.04 | 0.01 | 0.01 | 0.02 | 0.18 | 2.7 |
| | 1987 | 6.51 | 6.31 | 0.12 | 0.03 | 0.02 | 0 | 0.02 | 0.2 | 3.0 |
| | 1992 | 7.03 | 6.86 | 0.12 | 0.03 | 0.01 | 0 | 0.01 | 0.17 | 2.4 |
| | 1997 | 7.46 | 7.3 | 0.1 | 0.03 | 0.01 | 0 | 0.01 | 0.16 | 2.1 |
| Total nonprime cropland | 1982 | 76.91 | 39.68 | 13.92 | 7.35 | 4.4 | 2.85 | 8.71 | 37.23 | 48.4 |
| | 1987 | 73.23 | 37.73 | 13.73 | 7.21 | 4.32 | 2.75 | 7.49 | 35.5 | 48.5 |
| | 1992 | 67.27 | 40.2 | 11.66 | 5.51 | 3.11 | 1.88 | 4.91 | 27.07 | 40.2 |
| | 1997 | 66.62 | 42.07 | 11.51 | 5.03 | 2.69 | 1.52 | 3.81 | 24.55 | 36.9 |
| Nonprime cultivated | 1982 | 65.52 | 29.1 | 13.5 | 7.19 | 4.32 | 2.81 | 8.59 | 36.41 | 55.6 |
| | 1987 | 62.06 | 27.49 | 13.28 | 7.01 | 4.23 | 2.7 | 7.36 | 34.57 | 55.7 |
| | 1992 | 55.25 | 28.93 | 11.27 | 5.35 | 3.04 | 1.84 | 4.81 | 26.32 | 47.6 |
| | 1997 | 53.77 | 30.03 | 11.06 | 4.87 | 2.62 | 1.48 | 3.71 | 23.74 | 44.2 |
| Nonprime noncultivated | 1982 | 11.4 | 10.58 | 0.42 | 0.16 | 0.08 | 0.04 | 0.11 | 0.81 | 7.1 |
| | 1987 | 11.17 | 10.24 | 0.45 | 0.2 | 0.1 | 0.05 | 0.13 | 0.93 | 8.3 |
| | 1992 | 12.02 | 11.27 | 0.39 | 0.15 | 0.07 | 0.04 | 0.09 | 0.75 | 6.2 |
| | 1997 | 12.85 | 12.04 | 0.44 | 0.16 | 0.07 | 0.04 | 0.1 | 0.81 | 6.3 |

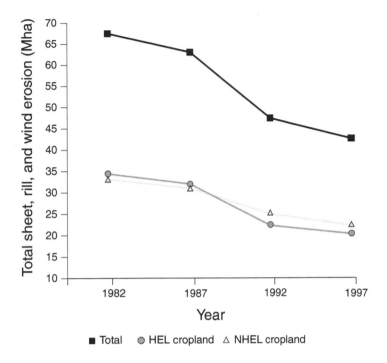

**Figure 8.1**   The extent of land area affected by severe (>T) sheet, rill, and wind erosion on total cropland.

The extent of erosion on prime cropland was 31.2 Mha (33.4%) in 1982, 28.6 Mha (31.3%) in 198, 21.2 Mha (24.3%) in 1992, and 19.0 Mha (22.2%) in 1997 (Figure 8.3). The extent of erosion on total prime cropland decreased over the 15-year period at an average rate of 0.8 Mha/yr. A large proportion of the decline in the extent of erosion occurred on prime cultivated land. Similar to the total prime cropland, the extent of erosion also declined on total nonprime cropland over the 15-year period. For example, the extent of erosion was 37.2 Mha (48.4%) in 1982, 35.5 Mha (48.5%) in 1987, 27.1 Mha (40.2%) in 1992, and 24.6 Mha (36.9%) in 1997. The mean rate of decline in the extent of erosion over this period was 0.8 Mha/yr or 2.7%/yr. Similar to the trends in total prime cropland, most of the decline in erosion also occurred in cultivated nonprime cropland (Table 8.3). These data show the importance of targeting the nonprime cropland for reducing the extent of erosion and minimizing the adverse off-site impact.

## 8.2 THE MAGNITUDE OF SHEET, RILL, AND WIND EROSION

In contrast to the area of the land, the magnitude of soil erosion is expressed in terms of the quantity (weight) of soil displaced. The weight of soil displaced by erosion is generally in the order cultivated HEL cropland > cultivated nonprime cropland > noncultivated HEL > noncultivated nonprime cropland.

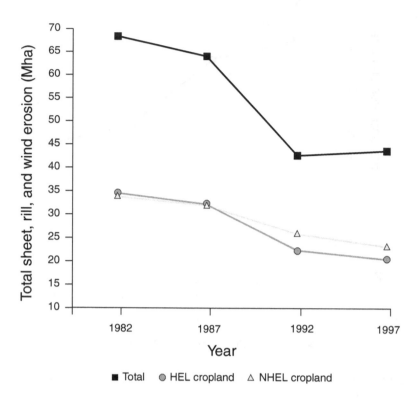

**Figure 8.2**   The extent of land area affected by severe (>T) sheet, rill, and wind erosion on cultivated cropland.

## 8.2.1   Total Cropland

The data in Table 8.5 show the magnitude of erosion on total cropland, further subdivided into HEL and NHEL categories. Total erosion, water and wind, on total cropland (comprising all categories >T) was 2.78 billion Mg in 1982, 2.54 billion Mg in 1987, 1.92 billion Mg in 1992, and 1.71 billion Mg in 1997. Similar to the water erosion, total erosion also decreased over the 15-year period. The average rate of decline in the magnitude of total erosion on total cropland was 71 million Mg/yr or 2.6%/yr (Figure 8.4). A larger fraction of the total cropland erosion occurred on HEL compared with the NHEL category. Nonetheless, total erosion declined on both categories of land. The mean rate of decline was 46.5 million Mg/yr or 3.0%/yr for HEL and 24.5 million Mg or 2.0%/yr for NHEL category.

## 8.2.2   Cultivated Cropland

The magnitude of erosion on total cropland presented in the previous section is divided into cultivated cropland (Table 8.6) and noncultivated cropland (Table 8.7). The magnitude of erosion on cultivated cropland was 2.74 billion Mg in 1982, 2.49 billion Mg in 1987, 1.89 billion Mg in 1992, and 1.67 billion Mg in 1997. The relative

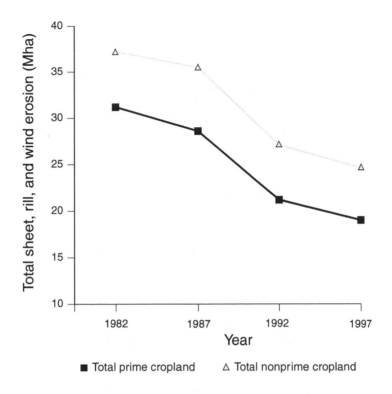

**Figure 8.3**  Temporal changes in the extent of sheet, rill, and wind erosion on prime and nonprime cropland.

magnitude of erosion expressed as a ratio of erosion from total cultivated cropland to total cropland was 98.5% in 1982, 98.2% in 1987, 98.0% in 1992, and 97.7% in 1997. Therefore, about 98% of the total erosion on cropland occurs on cultivated cropland. Similar to total cropland, the erosion from total cultivated land also decreased with time. The mean rate of decrease over the 15-year period was 71 million Mg/yr or 2.6%/yr. The magnitude of erosion also decreased on HEL and NHEL cultivated croplands. The mean rate of decline in the magnitude of erosion was 46.3 million Mg/yr or 3.0%/yr for HEL cultivated land compared with 24.5 million Mg/yr or 2.2%/yr for NHEL cultivated land (Table 8.6). The ratio of erosion from cultivated HEL to cultivated NHEL category was 1.3 for 1982, 1.2 for 1987, 1.14 for 1992, and 1.03 for 1997. The ratio of erosion also declined over time from 1.30 in 1982 to 1.03 in 1997 indicating the favorable impact of adopting BMPs on HEL for erosion control. The decrease may partly be attributed to conversion of the most susceptible land to that under CRP.

The data on total erosion on noncultivated land (categorized into total, HEL, and NHEL) are shown in Table 8.7. Total erosion from noncultivated land was small compared with that on cultivated land. Total erosion from total noncultivated land (Table 8.7), expressed as a fraction of total erosion from total cultivated land (Table 8.6), was 1.6% in 1982, 1.8% in 1987, 2.0% in 1992, and 2.3% in 1997. Erosion from

**Table 8.5  The Magnitude of Sheet, Rill, and Wind Erosion on Total Cropland in Terms of Severity Classes**

| Land Use | Year | ≤T | 1–2T | 2–3T | 3–4T | 4–5T | >5T | Total Magnitude >T |
|---|---|---|---|---|---|---|---|---|
| | | | | | 10⁶ Mg | | | |
| Total cropland | 1982 | 534.61 | 533.85 | 385.49 | 263.99 | 186.68 | 874.26 | 2778.88 |
| | 1987 | 510.57 | 516.76 | 357.69 | 255.29 | 180.48 | 715.69 | 2536.49 |
| | 1992 | 505.12 | 404.6 | 261.2 | 175.21 | 114.35 | 462.57 | 1923.06 |
| | 1997 | 495.0 | 376.42 | 229.48 | 151.59 | 96.47 | 363.59 | 1712.55 |
| Total HEL | 1982 | 72.12 | 140.34 | 176.61 | 176.67 | 153.98 | 854.61 | 1574.33 |
| | 1987 | 66.85 | 141.4 | 177.59 | 169.75 | 141.98 | 696.51 | 1394.08 |
| | 1992 | 69.99 | 135.72 | 151.82 | 127.52 | 93.57 | 454.46 | 1033.09 |
| | 1997 | 79.84 | 136.75 | 130.23 | 104.8 | 74.48 | 350.15 | 876.25 |
| Total NHEL | 1982 | 462.49 | 393.51 | 208.87 | 87.32 | 32.7 | 19.65 | 1204.54 |
| | 1987 | 443.72 | 375.35 | 180.1 | 85.54 | 38.5 | 19.18 | 1142.4 |
| | 1992 | 435.14 | 268.88 | 109.38 | 47.69 | 20.78 | 8.1 | 889.97 |
| | 1997 | 415.16 | 239.66 | 99.26 | 46.78 | 21.99 | 13.44 | 836.3 |

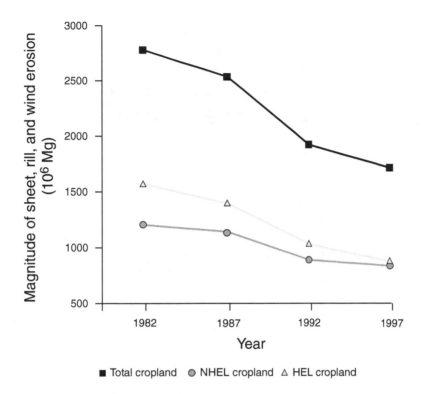

Figure 8.4    The magnitude of sheet, rill, and wind erosion on total cropland.

noncultivated land (for all three categories including total, HEL, and NHEL) increased in 1987 compared to 1982, and then decreased again in 1992 and 1997.

### 8.2.3    Prime Land

The magnitude of sheet, rill, and wind erosion on prime land (cultivated and noncultivated) is shown by the data in Table 8.8. The magnitude of erosion on total prime cropland was 1088 million Mg in 1982, 1009 million Mg in 1987, 802 million Mg in 1992, and 738 million Mg in 1997. The ratio of erosion from total nonprime land to that from total prime land was 1.55 in 1982, 1.51 in 1987, 1.40 in 1992, and 1.32 in 1997. The average rate of decline of erosion between 1982 and 1997 was 23.3 million Mg/yr or 2.1%/yr for total prime cropland compared with 47.8 million Mg/yr or 2.8%/yr for total nonprime cropland. As one would expect, a large proportion of erosion from nonprime cropland occurred on cultivated land. The fraction of erosion from cultivated nonprime cropland to that from total nonprime cropland was 98.1% in 1982, 97.8% in 1987, 97.4% in 1992, and 96.9% in 1997. In terms of the public policy, therefore, the target land for effective erosion/sedimentation control should be cultivated nonprime cropland. Conversion of HEL to CRP has been very effective in reducing erosion and transport of sediments in U.S. rivers.

Table 8.6 The Magnitude of Sheet, Rill, and Wind Erosion on Cultivated Cropland for Different Severity Classes

| Land Use | Year | ≤T | 1–2T | 2–3T | 3–4T | 4–5T | >5T | Total Magnitude >T |
|---|---|---|---|---|---|---|---|---|
| | | | | | $10^6$ Mg | | | |
| Total cultivated | 1982 | 517.31 | 527.49 | 391.28 | 261.25 | 184.77 | 863.83 | 2735.93 |
| | 1987 | 493.31 | 509.91 | 352.85 | 252.07 | 178.39 | 704.62 | 2491.14 |
| | 1992 | 487.61 | 398.52 | 257.59 | 173.01 | 112.88 | 455.59 | 1885.2 |
| | 1997 | 475.91 | 370.23 | 225.62 | 149.81 | 95.03 | 357.11 | 1673.7 |
| Cultivated HEL | 1982 | 62.79 | 135.36 | 173.24 | 174.32 | 152.29 | 844.4 | 1542.4 |
| | 1987 | 57.74 | 136.38 | 173.46 | 167.01 | 14.14 | 685.56 | 1360.27 |
| | 1992 | 60.03 | 131.57 | 148.58 | 125.68 | 92.64 | 447.59 | 1006.08 |
| | 1997 | 68.6 | 131.91 | 126.97 | 103.16 | 73.09 | 343.76 | 847.48 |
| Cultivated NHEL | 1982 | 454.52 | 392.13 | 208.04 | 86.93 | 32.48 | 19.43 | 1193.53 |
| | 1987 | 435.57 | 373.53 | 179.39 | 85.06 | 38.25 | 19.06 | 1130.86 |
| | 1992 | 427.58 | 266.95 | 109.01 | 47.34 | 20.25 | 8.0 | 879.12 |
| | 1997 | 407.31 | 238.32 | 98.65 | 46.65 | 21.94 | 13.35 | 826.22 |

Table 8.7 The Magnitude of Sheet, Rill, and Wind Erosion on Noncultivated Cropland in Terms of the Severity Classes

| Land Use | Year | ≤T | 1–2T | 2–3T | 3–4T | 4–5T | >5T | Total Magnitude |
|---|---|---|---|---|---|---|---|---|
| | | | | | $10^6$ Mg | | | |
| Total noncultivated | 1982 | 17.3 | 6.36 | 4.2 | 2.74 | 1.91 | 10.43 | 42.94 |
| | 1987 | 17.26 | 6.85 | 4.85 | 3.23 | 2.09 | 11.07 | 45.35 |
| | 1992 | 17.52 | 6.08 | 3.61 | 2.2 | 1.47 | 6.98 | 37.85 |
| | 1997 | 19.1 | 6.19 | 3.87 | 1.78 | 1.44 | 6.48 | 38.85 |
| Noncultivated HEL | 1982 | 9.33 | 4.98 | 3.37 | 2.36 | 1.68 | 10.21 | 31.93 |
| | 1987 | 9.11 | 5.03 | 4.14 | 2.74 | 1.85 | 10.95 | 33.81 |
| | 1992 | 9.96 | 4.15 | 3.24 | 1.84 | 0.94 | 6.88 | 27.0 |
| | 1997 | 11.24 | 4.84 | 3.26 | 1.64 | 1.39 | 6.38 | 28.76 |
| Noncultivated NHEL | 1982 | 7.97 | 1.38 | 0.83 | 0.39 | 0.22 | 0.22 | 11.02 |
| | 1987 | 8.15 | 1.82 | 0.71 | 0.48 | 0.25 | 0.13 | 11.54 |
| | 1992 | 7.56 | 1.93 | 0.38 | 0.35 | 0.53 | 0.1 | 10.85 |
| | 1997 | 7.85 | 1.34 | 0.6 | 0.14 | 0.06 | 0.1 | 10.09 |

Table 8.8 The Magnitude of Sheet, Rill, and Wind Erosion on Prime Cropland in Terms of the Severity Classes

| Land Use | Year | ≤T | 1–2T | 2–3T | 3–4T | 4–5T | >5T | Total Magnitude |
|---|---|---|---|---|---|---|---|---|
| | | | | | 10⁶ Mg | | | |
| Total prime cropland | 1982 | 348.71 | 296.6 | 184.7 | 101.53 | 55.26 | 100.44 | 1087.25 |
| | 1987 | 337.6 | 281.19 | 158.65 | 95.81 | 55.24 | 80.84 | 1009.34 |
| | 1992 | 332.32 | 209.6 | 114.24 | 62.42 | 30.23 | 53.21 | 802.04 |
| | 1997 | 321.19 | 189.11 | 95.56 | 56.86 | 29.92 | 42.63 | 738.27 |
| Prime cultivated | 1982 | 343.27 | 295.16 | 183.8 | 101.18 | 54.9 | 97.88 | 1076.19 |
| | 1987 | 332.2 | 279.49 | 157.85 | 95.26 | 55.06 | 78.39 | 998.27 |
| | 1992 | 327.03 | 208.03 | 113.6 | 62.07 | 30.08 | 52.46 | 793.27 |
| | 1997 | 315.36 | 187.81 | 97.84 | 56.68 | 29.76 | 41.8 | 729.25 |
| Prime noncultivated | 1982 | 5.44 | 1.44 | 0.91 | 0.35 | 0.36 | 2.56 | 11.06 |
| | 1987 | 5.4 | 1.7 | 0.8 | 0.55 | 0.18 | 2.45 | 11.07 |
| | 1992 | 5.29 | 1.58 | 0.64 | 0.36 | 0.15 | 0.75 | 8.77 |
| | 1997 | 5.83 | 1.3 | 0.72 | 0.18 | 0.16 | 0.84 | 9.03 |
| Total nonprime cropland | 1982 | 185.87 | 237.2 | 200.78 | 162.46 | 131.42 | 773.82 | 1691.54 |
| | 1987 | 172.94 | 235.49 | 199.04 | 159.48 | 125.24 | 634.85 | 1527.04 |
| | 1992 | 172.68 | 197.95 | 146.96 | 112.77 | 84.11 | 409.02 | 1120.5 |
| | 1997 | 173.71 | 187.26 | 130.93 | 94.72 | 66.53 | 320.62 | 973.76 |
| Nonprime cultivated | 1982 | 174.01 | 232.28 | 197.49 | 160.07 | 129.87 | 765.95 | 1659.66 |
| | 1987 | 161.09 | 230.33 | 194.99 | 156.8 | 123.32 | 626.23 | 1492.76 |
| | 1992 | 160.46 | 190.45 | 143.99 | 110.93 | 82.79 | 402.78 | 1091.41 |
| | 1997 | 160.44 | 182.37 | 127.78 | 93.12 | 65.25 | 314.97 | 943.94 |
| Nonprime noncultivated | 1982 | 11.86 | 4.92 | 3.3 | 2.39 | 1.54 | 7.87 | 31.88 |
| | 1987 | 11.86 | 5.16 | 4.05 | 2.68 | 1.92 | 8.62 | 34.28 |
| | 1992 | 12.22 | 4.5 | 2.97 | 1.84 | 1.32 | 6.23 | 29.09 |
| | 1997 | 13.27 | 4.88 | 3.15 | 1.6 | 1.28 | 5.64 | 29.82 |

## 8.3 THE RATE OF SHEET, RILL, AND WIND EROSION

The rate of erosion is expressed in terms of the magnitude (quantity) of soil displaced per unit area and per unit times (Mg/ha/yr). The severity of erosion rate can be judged when expressed in terms of the T value. Therefore, the data in Table 8.9 to Table 8.12 and in Figure 8.4 show erosion rate on cropland in relation to different land uses and susceptibility categories. The rate of erosion increases with increase in severity classes from ≤T to >5T. The average rate of erosion was in the order HEL > NHEL for total, cultivated, and noncultivated land (Figure 8.5). Similarly, the rate of erosion was nonprime cropland then prime cultivated for both cultivated and noncultivated categories.

**Table 8.9   Distribution of Total Cropland for Sheet, Rill, and Wind Erosion in Terms of the Severity Classes**

| Land Use | Year | ≤T | 1–2T | 2–3T | 3–4T | 4–5T | >5T | Average Rate |
|---|---|---|---|---|---|---|---|---|
| | | | | | Mg/ha | | | |
| Total cropland | 1982 | 5.2 | 16.4 | 26.9 | 36.6 | 45.2 | 85.5 | 16.3 |
| | 1987 | 5.1 | 16.5 | 27.0 | 36.7 | 45.4 | 82.2 | 15.4 |
| | 1992 | 4.7 | 16.0 | 26.4 | 36.0 | 44.3 | 80.9 | 12.4 |
| | 1997 | 4.5 | 15.7 | 26.1 | 35.7 | 44.3 | 81.6 | 11.2 |
| Total HEL | 1982 | 4.4 | 16.2 | 25.8 | 35.3 | 44.2 | 86.0 | 31.2 |
| | 1987 | 4.4 | 16.3 | 26.0 | 35.2 | 44.2 | 82.6 | 29.5 |
| | 1992 | 4.2 | 16.1 | 25.5 | 34.7 | 43.3 | 81.3 | 24.3 |
| | 1997 | 4.2 | 15.5 | 24.7 | 33.9 | 42.8 | 82.2 | 20.8 |
| Total NHEL | 1982 | 5.4 | 16.4 | 27.9 | 39.6 | 50.8 | 68.3 | 10.0 |
| | 1987 | 5.2 | 16.6 | 28.0 | 40.2 | 50.7 | 69.7 | 9.7 |
| | 1992 | 4.8 | 16.0 | 27.7 | 40.0 | 49.2 | 64.7 | 7.9 |
| | 1997 | 4.6 | 15.8 | 28.2 | 40.6 | 50.7 | 68.9 | 7.6 |

**Table 8.10   Distribution of Cultivated Cropland for Sheet, Rill, and Wind Erosion in Terms of the Severity Classes**

| Land Use U.S. Total | Year | ≤T | 1–2T | 2–3T | 3–4T | 4–5T | >5T | Average Rate |
|---|---|---|---|---|---|---|---|---|
| | | | | | Mg/ha | | | |
| Total cultivated | 1982 | 6.1 | 16.4 | 26.9 | 36.7 | 45.3 | 85.5 | 17.9 |
| | 1987 | 5.9 | 16.6 | 27.1 | 36.8 | 45.6 | 82.3 | 16.9 |
| | 1992 | 5.5 | 16.1 | 26.5 | 36.2 | 44.4 | 81.1 | 13.9 |
| | 1997 | 5.3 | 15.8 | 26.2 | 35.9 | 44.5 | 82.1 | 12.6 |
| Cultivated HEL | 1982 | 6.1 | 16.5 | 25.9 | 35.4 | 44.2 | 86.0 | 35.5 |
| | 1987 | 6.1 | 16.6 | 26.2 | 35.3 | 44.3 | 82.7 | 33.6 |
| | 1992 | 6.0 | 16.3 | 25.7 | 34.9 | 43.5 | 81.5 | 28.7 |
| | 1997 | 6.0 | 15.7 | 24.9 | 34.2 | 43.0 | 82.7 | 25.0 |
| Cultivated NHEL | 1982 | 6.1 | 16.4 | 27.9 | 39.7 | 50.8 | 68.2 | 10.9 |
| | 1987 | 5.8 | 16.6 | 28.0 | 40.2 | 50.7 | 69.7 | 10.6 |
| | 1992 | 5.5 | 16.0 | 27.7 | 40.1 | 49.2 | 64.6 | 8.7 |
| | 1997 | 5.2 | 15.8 | 28.2 | 40.6 | 50.8 | 68.9 | 8.4 |

**Table 8.11**   **Distribution of Noncultivated Cropland for Sheet, Rill, and Wind Erosion in Terms of the Severity Classes**

| Land Use | Year | ≤T | 1–2T | 2–3T | 3–4T | 4–5T | >5T | Average Rate |
|---|---|---|---|---|---|---|---|---|
| | | | | Mg/ha | | | | |
| Total noncultivated | 1982 | 1 | 12.1 | 21.3 | 29.9 | 39.3 | 80.8 | 2.4 |
| | 1987 | 1 | 11.9 | 21.1 | 27.9 | 35.8 | 74.3 | 2.6 |
| | 1992 | 1 | 11.9 | 19.5 | 27.2 | 35.4 | 68.2 | 2.0 |
| | 1997 | 1 | 11.3 | 19.5 | 25.0 | 34.0 | 59.6 | 1.9 |
| Noncultivated HEL | 1982 | 1.5 | 11.6 | 19.9 | 29.3 | 38.1 | 80.8 | 4.5 |
| | 1987 | 1.6 | 11.4 | 20.4 | 26.8 | 34.4 | 74.6 | 5.0 |
| | 1992 | 1.5 | 11.1 | 19.0 | 26.5 | 31.0 | 68.2 | 3.6 |
| | 1997 | 1.5 | 10.8 | 18.5 | 24.3 | 33.8 | 59.4 | 3.5 |
| Noncultivated NHEL | 1982 | 0.7 | 14.1 | 29.8 | 34.4 | 0 | 0 | 1.0 |
| | 1987 | 0.8 | 13.8 | 25.9 | 36.0 | 50.0 | 0 | 1.1 |
| | 1992 | 0.7 | 14.0 | 24.4 | 31.2 | 47.2 | 0 | 0.9 |
| | 1997 | 0.7 | 13.5 | 27.6 | 35.8 | 38.9 | 77 | 0.8 |

**Table 8.12**   **Distribution of Prime Cropland for Sheet, Rill, and Wind Erosion in Terms of the Severity Classes**

| Land Use | Year | ≤T | 1–2T | 2–3T | 3–4T | 4–5T | >5T | Average Rate |
|---|---|---|---|---|---|---|---|---|
| | | | | Mg/ha | | | | |
| Total prime cropland | 1982 | 5.6 | 15.9 | 26.4 | 36.3 | 43.2 | 66.3 | 11.6 |
| | 1987 | 5.4 | 16.1 | 26.3 | 36.4 | 45.2 | 66.5 | 11.0 |
| | 1992 | 5.0 | 15.4 | 26.0 | 35.7 | 43.1 | 66.1 | 9.2 |
| | 1997 | 4.8 | 15.2 | 26.2 | 36.7 | 45.5 | 66.4 | 8.6 |
| Prime cultivated | 1982 | 6.2 | 15.9 | 26.4 | 36.3 | 43.2 | 65.4 | 12.4 |
| | 1987 | 5.9 | 16.1 | 26.3 | 36.5 | 45.2 | 65.5 | 11.8 |
| | 1992 | 5.5 | 15.4 | 26.1 | 35.7 | 43.2 | 66.1 | 9.9 |
| | 1997 | 5.3 | 15.2 | 26.2 | 36.7 | 45.6 | 66.4 | 9.3 |
| Prime noncultivated | 1982 | 0.8 | 13.4 | 25.4 | 36.1 | 45.2 | 142.8 | 1.7 |
| | 1987 | 0.9 | 13.8 | 23.2 | 33.1 | 39.8 | 124.7 | 1.7 |
| | 1992 | 0.8 | 13.6 | 20.8 | 33.6 | 38.0 | 69.0 | 1.2 |
| | 1997 | 0.8 | 12.6 | 20.7 | 32.8 | 40.8 | 7.3 | 1.2 |
| Total nonprime cropland | 1982 | 4.7 | 17.0 | 27.3 | 36.9 | 46.1 | 88.8 | 22.0 |
| | 1987 | 4.6 | 17.1 | 27.6 | 36.8 | 45.5 | 84.7 | 20.8 |
| | 1992 | 4.3 | 16.7 | 26.7 | 36.2 | 44.7 | 83.3 | 16.6 |
| | 1997 | 4.1 | 16.3 | 26.0 | 35.2 | 43.8 | 84.1 | 14.6 |
| Nonprime cultivated | 1982 | 6.0 | 17.2 | 27.4 | 37.0 | 46.2 | 89.0 | 25.3 |
| | 1987 | 5.9 | 17.3 | 27.8 | 37.1 | 45.7 | 85.0 | 24.0 |
| | 1992 | 5.5 | 16.9 | 26.9 | 36.4 | 44.9 | 83.6 | 19.7 |
| | 1997 | 5.3 | 16.5 | 26.2 | 35.5 | 44.1 | 84.8 | 17.5 |
| Nonprime noncultivated | 1982 | 1.1 | 11.8 | 20.4 | 29.2 | 38.2 | 7.8 | 2.8 |
| | 1987 | 1.2 | 11.4 | 20.7 | 27.0 | 35.4 | 66.7 | 3.1 |
| | 1992 | 1.1 | 11.4 | 19.2 | 26.2 | 35.1 | 68.2 | 2.4 |
| | 1997 | 1.1 | 11.0 | 19.2 | 24.3 | 33.3 | 58.3 | 2.3 |

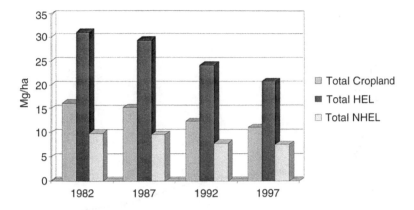

**Figure 8.5**   The average rate of erosion for total cropland, total HEL, and total NHEL.

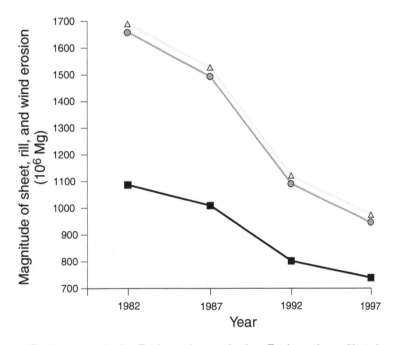

■ Total prime cropland  ● Total nonprime cropland  △ Total nonprime cultivated

**Figure 8.6**   The magnitude of sheet, rill, and wind erosion on prime and nonprime cropland.

## 8.3.1   Total Cropland

The rate of erosion on HEL cropland of all categories was 31.2 Mg/ha/yr in 1982, 29.5 Mg/ha/yr in 1987, 24.3 Mg/ha/yr in 1992, and 20.8 Mg/ha/yr in 1997 (Table 8.9). The rate of erosion on the NHEL category was about one-third of that on HEL. The rate of erosion also decreased with time in all categories of land (Figure 8.6 to Figure 8.9).

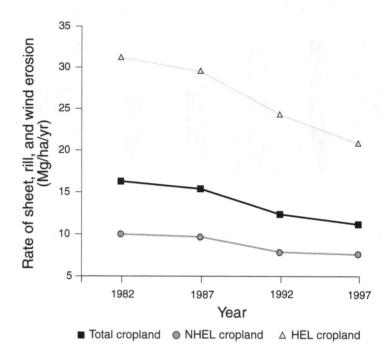

■ Total cropland   ● NHEL cropland   △ HEL cropland

**Figure 8.7**   The rate of erosion on HEL and NHEL cropland.

## 8.3.2   Cultivated Cropland

Similar to total cropland, the rate of erosion was higher on cultivated HEL than on NHEL (Table 8.10). The erosion rate on NHEL was 30.7% of that on NHEL in 1982, 31.5% in 1987, 30.3% in 1992, and 33.6% in 1997. Compared to cultivated cropland, the rate of erosion was drastically less on noncultivated cropland for both HEL and NHEL categories (Table 8.11). Cultivation exacerbates the rate of soil erosion. Total erosion rate on cultivated HEL (Table 8.1) compared to that on noncultivated HEL (Table 8.11) was 7.9 times in 1982, 6.7 times in 1987, 8.0 times in 1992, and 8.2 times in 1997. The erosion rate decreased over time in all land use categories.

## 8.3.3   Prime Land

The rate of erosion was more on nonprime cultivated cropland than prime cultivated cropland (Table 8.12). In general, the rate declined over time. The rate of erosion from nonprime cultivated land was about twice that from prime cultivated cropland. Once again, the target for erosion control measures should be the nonprime cultivated land.

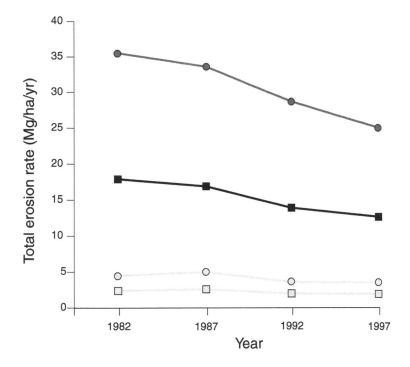

**Figure 8.8** Rate of sheet, rill, and wind erosion on cultivated HEL, total cultivated, noncultivated HEL, and total noncultivated lands. (See Table 8.10 and Table 8.11 for
details.)

## 8.4 CONCLUSIONS

The extent, magnitude, and rate of wind and water erosion were greater on
cultivated than noncultivated cropland. The erosion hazard was the least on prime
cropland and the most on nonprime cultivated land. Total erosion, the sum of wind
and water erosion on cropland, also decreased over the 15-year period (Table 8.13
and Table 8.14). The relative reduction in erosion hazard over time was the most on
land use and categories that were most susceptible to soil erosion, e.g., HEL-
cultivated land and nonprime cultivated land. This observation is consistent with the
strategy that soil erosion hazard depends on how soil is managed.

The data in Table 8.13 show that the rate of decline in the extent of severe (>T)
erosion on all cropland (cultivated and uncultivated) over the 15-year period from
1982 to 1997 was 1.91 Mha/yr for total cropland, 0.76 Mha/yr for all HEL cropland,
and 1.04 Mha/yr for NHEL cropland. The reduction in severe (>T) erosion for
cultivated cropland was 1.81 Mha/yr for all cultivated cropland, 0.77 Mha/yr for
HEL cropland, and 1.04 Mha/yr for NHEL cropland. The reduction in extent of

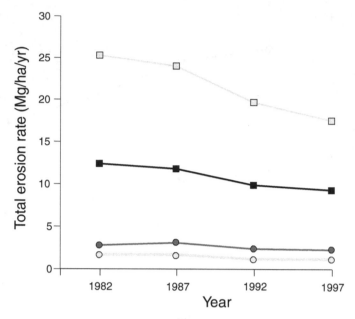

■ Prime cultivated □ Nonprime cultivated ○ Prime noncultivated
● Nonprime noncultivated

**Figure 8.9**   Rate of sheet, rill, and wind erosion on cultivated and noncultivated prime land.
(See Table 8.12 for details.)

**Table 8.13**   **Regression Equations Relating the Extent of Total Sheet, Rill, and Wind Erosion with Time between 1982 (Base Year) and 1997 (15 Years)**

| Dependent Variable | Regression Equation | $R^2$ |
|---|---|---|
| I. Severe erosion (>T) on all cropland (Mha) | | |
|   (i)  Total cropland | $Y = 69.0 - 1.91x$ | 0.94 |
|   (ii)  HEL cropland | $Y = 34.4 - 0.76x$ | 0.97 |
|   (iii)  NHEL cropland | $Y = 35.2 - 1.04x$ | 0.92 |
| II. Severe erosion (>T) on cultivated cropland (Mha) | | |
|   (i)  Total cropland | $Y = 68.7 - 1.81x$ | 0.84 |
|   (ii)  HEL cropland | $Y = 33.6 - 0.77x$ | 0.96 |
|   (iii)  NHEL cropland | $Y = 35.1 - 1.04x$ | 0.92 |
| III. Prime land (Mha) | | |
|   (i)  Total prime land | $Y = 31.6 - 0.88x$ | 0.95 |
|   (ii)  Total nonprime cropland | $Y = 38.0 - 0.92x$ | 0.93 |

X = period in years. For example, x is 0 for 1982 and x is 15 for 1997.

severe (>T) erosion was 0.88 Mha/yr on total prime land and 0.92 Mha/yr on nonprime cropland.

The data in Table 8.14 show these trends of decline in the magnitude and rate of erosion over the 15-year period from 1982 to 1997. The magnitude of erosion declined at the rate of $76.2 \times 10^6$ Mg/yr for NHEL cropland. There were similar

**Table 8.14   Regression Equations during the Magnitude and Rate of Sheet, Rill, and Wind Erosion with Time between 1982 (Base Year) and 1997 (15 Years) on U.S. Cropland**

| Dependent Variable | Regression Equation | $R^2$ |
|---|---|---|
| I. Magnitude of erosion from all cropland ($10^6$ Mg) | | |
|   (i) Total cropland | Y = 2809 − 76.2x | 0.96 |
|   (ii) HEL cropland | Y = 1587 − 49.1x | 0.98 |
|   (iii) NHEL cropland | Y = 1222 − 27.2x | 0.93 |
| II. Magnitude of erosion from prime cropland ($10^6$ Mg) | | |
|   (i) Total prime cropland | Y = 1097 − 25.0x | 0.95 |
|   (ii) Total prime cultivated cropland | Y = 1712 − 51.2x | 0.96 |
|   (iii) Total nonprime cropland | Y = 1679 − 51.0x | 0.96 |
| III. Rate of erosion from erodible cropland (Mg/ha/yr) | | |
|   (i) Total cropland | Y = 16.6 − 0.37x | 0.96 |
|   (ii) HEL cropland | Y = 31.9 − 0.73x | 0.97 |
|   (iii) NHEL cropland | Y = 10.1 − 0.118x | 0.90 |
| IV. Rate of erosion from prime cropland (Mg/ha/yr) | | |
|   (i) Prime cultivated | Y = 12.5 − 0.224x | 0.95 |
|   (ii) Nonprime cultivated | Y = 25.8 − 0.554x | 0.96 |
|   (iii) Prime noncultivated | Y = 1.75 − 0.04x | 0.80 |

X = the period in years. For example, x is 0 for 1982 and x is 15 for 1997.

trends of decline in magnitude for erosion on prime and nonprime cropland. The magnitude of soil erosion declined at the rate of $25.0 \times 10^6$ Mg/yr for total prime cropland, $51.2 \times 10^6$ Mg/yr for total prime cultivated cropland, and $51.0 \times 10^6$ Mg/yr for total nonprime cropland. Similar trends were observed in the rate of soil erosion (Table 8.14).

There is an important policy implication of the data presented in Chapter 8. Since soil erosion is most severe on HEL and nonprime cultivated land, public policy must be directed to targeting these ecologically sensitive ecoregions. Conversion of HEL and nonprime land to other land uses (e.g., forestry, permanent cover, etc.), and adoption of BMPs on NHEL and prime cropland are important strategies toward sustainable management of soil and water resources.

Erosion control and its management are also effective strategies to improve water and air quality and mitigate the greenhouse effect. Environmental impacts of erosion control will be discussed in the concluding chapter of this book.

## REFERENCES

Crosson, P. and J.R. Anderson. Land degradation and food security: economic impacts of watershed degradation. In *Integrated Watershed Management in the Global Ecosystem*. R. Lal (Ed.), CRC Press, Boca Raton, FL, 2000, pp. 291–303.

den Biggelaar, C., R. Lal, K. Wiebe, and V. Breneman. Impact of soil erosion on crop yields in North America. *Adv. Agron.* 72, 1–52, 2001.

Foster, G. and S. Dabney. *Agricultural Tillage Systems: Water Erosion and Sedimentation, Farming for a Better Environment*. Soil Water Conserv. Soc., Ankeny, IA, 1995.

Lal, R. Soil erosion impact on agronomic productivity and environment quality. *CRC Crit. Rev. Plant Sci.* 17, 319–464, 1998.

National Research Council. *Soil and Water Quality.* National Academy Press, Washington, D.C., 1993.

National Resources Conservation Service. *A Geography of Hope.* USDA-NRCS, Washington, D.C., 1995.

Piper, S. and L. Lee. Estimating the off-site household damage from wind erosion in the western United States. Staff Report 8929. USDA-ERS, Washington, D.C., 1989.

USEPA. National water quality inventory: office of water. 1994 report to Congress. Washington, D.C., 1995.

# CHAPTER 9

# Total Erosion, Wind Erosion, and Erosion on Pasture and CRP Lands

The severity of soil erosion, both water and wind, on pasture and Conservation Reserve Program (CRP) lands (Plate 9.1) is less than that on cropland for two reasons: (1) more vegetation cover, and (2) less soil disturbance due to tillage and farm/vehicular traffic. High vegetal cover and low frequency of soil disturbance enhance the soil organic carbon (SOC) pool, improve soil structure, and increase biotic activity of soil fauna (e.g., earthworms). Improvement in soil physical quality enhances structural stability and decreases soil erodibility, increases water infiltration capacity, decreases runoff, and reduces the overall risks of soil erosion. Herein lies the principle of erosion control, which involves conversion of highly erodible land (HEL) and nonprime cropland to pastures, CRP, or perennial crops. With progressive increase in vegetal/ground cover over time after conversion from cropland to pasture, perennial crops, or CRP, the severity of erosion will progressively decrease. Well-managed pasture and CRP lands have drastically less erosion hazard than croplands. On the other hand, livestock trails up and down the hill can cause severe gullying on even well-managed pastureland (Plate 9.2).

## 9.1 TOTAL EROSION ON PASTURELAND

The data in Table 9.1 to Table 9.3, and Figure 9.1 show the extent, magnitude, and rate of total erosion on U.S. pastureland. The area of pastureland in terms of different severity classes (T values) is depicted in Table 9.1. The total area of pastureland with severity of erosion >T was 4.2 Mha (7.9%) in 1992 and 3.2 Mha (6.5%) in 1997. The land area affected by erosion decreased over time in all erosion severity classes. The average rate of decline in land area over the 15-year period with erosion >T was 0.07 Mha/yr. The mean rate of decline was 0.25 Mha/yr or 0.5%/yr for ≤T class, 0.024 Mha/yr or 1.1%/yr for 1–2T, 0.017 Mha/yr of 1.9%/yr for 2–3T, 0.011 Mha/yr or 2.6%/yr for 3-4T, 0.005 Mha/yr or 2.5%/yr for 4–5T, and 0.013 Mha/yr or 2.7%/yr for >5T class of soil erosion. In contrast, the decline in land area under pasture for

**Plate 9.1** Establishment of vegetal cover on CRP land decreases risks of soil erosion.

the same period was 0.6%/yr. Therefore, the rate of decline in the area affected by severe erosion was more than that of the change in area under pasture.

The magnitude of water and wind erosion from pastureland is shown in Table 9.2. The amount of soil displaced by both water and wind erosion on pastureland was 138.2 million Mg in 1982, 126 million Mg in 198, 120 million Mg in 1992, and 104 million Mg in 1997.

The rate of soil erosion (Mg/ha) on pastureland for the 15-year period for different severity classes is shown in Table 9.3. There was a decline in the rate in all erosion severity classes. From 1982 to 1987, the rate declined from 1.2 to 1.1 Mg/ha in ≤T class, 10.8 to 10.2 Mg/ha in 1–2T, 18.1 to 17.2 Mg/ha in 2–3T, 23.8 to 21.9 Mg/ha in 3–4T, 29.4 to 27.5 Mg/ha in 4–5T, 52.2 to 49.9 Mg/ha in >5T class, and 2.6 to 2.1 Mg/ha in total pastureland area (Table 9.3). The decline in the rate of erosion, over and above the decline in the land area, is due to adoption of best management practices (BMPs) on pastureland. These BMPs include growing improved species with high biomass production, and rapid and high ground cover establishment. Soil fertility management and controlled grazing would also reduce the risks of soil erosion. The decline in the extent (area affected) and magnitude (quantity of soil displaced) over the 15-year period from 1982 to 1997 is shown by Equation 9.1 and Equation 9.2.

$$Y(Mha) = 4.19 - 0.062X \qquad (9.1)$$

$$Y = (10^6 \text{ Mg}) = 138 - 2.16X \qquad (9.2)$$

where X is the number of years (i.e., 0 for 1982, 5 for 1987, 10 for 1992, and 15 for 1997).

**Plate 9.2**  An example of severe gullying on pastureland as a result of livestock traffic.

**Table 9.1  Extent of Pastureland Affected by Water and Wind Erosion in Terms of the Severity Classes**

| Year | Total Area | ≤T | 1–2T | 2–3T | 3–4T | 4–5T | >5T | Total Area >T | >T as % of Total Area |
|------|------------|-----|------|------|------|------|------|---------------|-----------------------|
| | | | | Mha | | | | | |
| 1982 | 53.4 | 49.2 | 2.21 | 0.89 | 0.44 | 0.21 | 0.47 | 4.22 | 7.9 |
| 1987 | 51.9 | 48.1 | 2.02 | 0.77 | 0.38 | 0.21 | 0.43 | 3.81 | 7.3 |
| 1992 | 51.0 | 47.3 | 2.04 | 0.73 | 0.38 | 0.20 | 0.37 | 3.72 | 7.3 |
| 1997 | 48.6 | 45.4 | 1.84 | 0.63 | 0.27 | 0.13 | 0.28 | 3.15 | 6.5 |

## 9.2 TOTAL EROSION ON CRP LAND

The CRP program emerged as a result of the Food Security Act of 1985. Some of the HEL was converted to permanent cover under CRP, with significant benefits of reduced soil erosion (Plate 9.1). The data on the extent of water and wind erosion (Mha) on CRP land under different severity classes is shown in Table 9.4 and Figure 9.2. Total area of CRP land with severity ≥T was 2.36 Mha (42.2%) in 1987,

Table 9.2    The Magnitude of Water and Wind Erosion on Total
             Pastureland in Terms of the Severity Classes

| Year | ≤T | 1–2T | 2–3T | 3–4T | 4–5T | >5T | Total |
|------|------|-------|-------|-------|------|-------|--------|
| | | | | $10^6$ Mg | | | |
| 1982 | 56.91 | 23.78 | 16.14 | 10.54 | 6.36 | 24.48 | 138.2 |
| 1987 | 54.16 | 21.35 | 13.64 | 8.56 | 6.18 | 22.27 | 126.16 |
| 1992 | 53.12 | 21.17 | 12.7 | 8.62 | 5.86 | 18.89 | 120.37 |
| 1997 | 51.29 | 18.78 | 10.85 | 5.85 | 3.73 | 13.72 | 104.22 |

Table 9.3    The Rate of Water and Wind Erosion on Pastureland
             in Terms of the Severity Classes

| Year | ≤T | 1–2T | 2–3T | 3–4T | 4–5T | >5T |
|------|------|-------|-------|-------|-------|-------|
| | | | Mg/ha | | | |
| 1982 | 1.2 | 10.8 | 18.1 | 23.8 | 29.4 | 52.2 |
| 1987 | 1.1 | 10.5 | 17.8 | 22.8 | 29.9 | 52.2 |
| 1992 | 1.1 | 10.4 | 17.3 | 22.7 | 28.8 | 51.1 |
| 1997 | 1.1 | 10.2 | 17.2 | 21.9 | 27.5 | 49.9 |

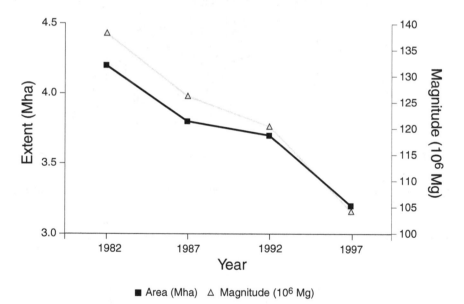

■ Area (Mha)   △ Magnitude ($10^6$ Mg)

Figure 9.1    The extent and magnitude of erosion on pastureland. (See Table 9.1 and Table
              9.2 for details.)

1.0 Mha (1.0%) in 1992, and 0.32 Mha (2.4%) in 1987. The mean rate of decline
in the total area affected by erosion on CRP land was 0.20 Mha/yr. Relative reduction
was more on HEL-CRP land than on NHEL-CRP land. Conversion of HEL cropland
to CRP drastically reduced the risks of soil erosion.

The magnitude of erosion (million Mg) on total CRP, further divided into HEL and
NHEL categories, is shown by the data in Table 9.5 and Figure 9.3. The magnitude

Table 9.4  The Extent of Water and Wind Erosion on CRP Land in Terms of the Severity Classes

| Land Use | Year | Total Area | ≤T | 1–2T | 2–3T | 3–4T | 4–5T | >5T | Total Area >T | >T as % of Total Area |
|---|---|---|---|---|---|---|---|---|---|---|
| | | | | | Mha | | | | | |
| Total CRP | 1987 | 5.59 | 3.23 | 0.79 | 0.48 | 0.26 | 0.23 | 2.36 | 2.36 | 42.2 |
| | 1992 | 13.78 | 13.00 | 0.44 | 0.33 | 0.07 | 0.04 | 1.00 | 1.00 | 7.3 |
| | 1997 | 13.24 | 12.93 | 0.17 | 0.05 | 0.02 | 0.03 | 0.32 | 0.32 | 2.4 |
| CRP-HEL | 1987 | 3.44 | 1.65 | 0.43 | 0.34 | 0.23 | 0.21 | 1.79 | 1.79 | 52.1 |
| | 1992 | 7.81 | 7.16 | 0.35 | 0.11 | 0.06 | 0.03 | 0.67 | 0.67 | 8.6 |
| | 1997 | 7.55 | 7.28 | 0.15 | 0.04 | 0.016 | 0.022 | 0.27 | 0.27 | 3.6 |
| CRP-NHEL | 1987 | 2.15 | 1.58 | 0.36 | 0.14 | 0.033 | 0.014 | 0.56 | 0.56 | 26.1 |
| | 1992 | 5.97 | 5.85 | 0.09 | 0.021 | 0.004 | 0 | 0.123 | 0.123 | 2.1 |
| | 1997 | 5.68 | 5.65 | 0.016 | 0.009 | 0.005 | 0 | 0.034 | 0.034 | 0.6 |

**Table 9.5    The Magnitude of Sheet, Rill, and Wind Erosion on CRP Land in Terms of the Severity Classes**

| Land Use | Year | ≤T | 1–2T | 2–3T | 3–4T | 4–5T | >5T | Total >T |
|---|---|---|---|---|---|---|---|---|
| | | | | | $10^6$ Mg | | | |
| Total CRP | 1987 | 9.38 | 13.01 | 13.11 | 9.09 | 10.16 | 55.91 | 110.67 |
| | 1992 | 16.05 | 6.58 | 2.71 | 2.16 | 1.48 | 9.94 | 38.92 |
| | 1997 | 10.87 | 2.46 | 1.02 | 0.57 | 0.76 | 4.69 | 20.38 |
| CRP-HEL | 1987 | 4.62 | 6.98 | 9.52 | 7.92 | 9.58 | 55.11 | 93.72 |
| | 1992 | 10.33 | 5.28 | 2.17 | 2.01 | 1.14 | 9.94 | 30.87 |
| | 1997 | 7.7 | 2.23 | 0.77 | 0.42 | 0.57 | 4.69 | 16.37 |
| CRP-NHEL | 1987 | 4.76 | 6.03 | 3.59 | 1.18 | 0.58 | 0.8 | 16.95 |
| | 1992 | 5.72 | 1.3 | 0.54 | 0.15 | 0.34 | 0 | 8.04 |
| | 1997 | 3.17 | 0.23 | 0.26 | 0.15 | 0.2 | 0 | 4.01 |

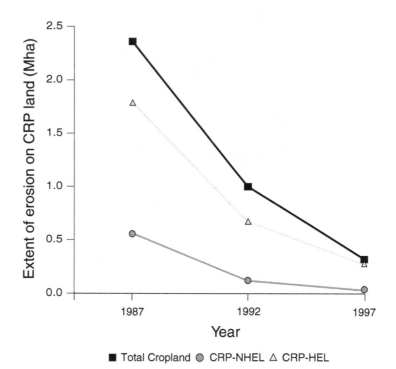

■ Total Cropland ● CRP-NHEL △ CRP-HEL

**Figure 9.2**    Extent of erosion on CRP land. (See Table 9.4 for details.)

of erosion in 1987, 1992, and 1997, respectively, was 110.7 million Mg, 38.9 million Mg, and 20.4 million Mg for total CRP; 93.7 million Mg, 30.9 million Mg, and 16.4 million Mg in 1992; and 17.0 million Mg, 8.0 million Mg, and 4.0 million Mg in 1997.

In accord with the magnitude, the rate of soil erosion on CRP land also decreased over time. The highest rate of erosion was observed on HEL-CRP and the lowest

Table 9.6    The Rate of Sheet, Rill, and Wind Erosion on CRP Land
             in Terms of the Severity Classes

| Land Use | Year | ≤T | 1–2T | 2–3T | 3–4T | 4–5T | >5T |
|----------|------|-----|------|------|------|------|------|
|          |      | Mg/ha | | | | | |
| Total CRP | 1987 | 2.9 | 16.4 | 27.4 | 34.6 | 45.3 | 93.5 |
|           | 1992 | 1.2 | 15.0 | 23.8 | 32.3 | 36.1 | 86.0 |
|           | 1997 | 0.8 | 14.6 | 22.3 | 29.9 | 28.9 | 99.5 |
| CRP-HEL | 1987 | 2.8 | 16.3 | 27.8 | 34.5 | 45.5 | 94.4 |
|         | 1992 | 1.4 | 15.2 | 23.6 | 32.3 | 33.9 | 86.0 |
|         | 1997 | 1.1 | 14.6 | 21.0 | 29.3 | 25.3 | 99.5 |
| CRP-NHEL | 1987 | 3.0 | 16.6 | 26.3 | 35.6 | 42.7 | 55.6 |
|          | 1992 | 1.0 | 14.3 | 25.0 | 32.5 | 45.5 | NA |
|          | 1997 | 0.6 | 14.2 | 27.5 | 31.8 | 49.1 | NA |

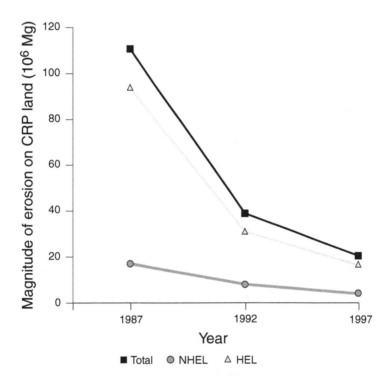

**Figure 9.3**    The magnitude of total erosion on CRP land. (see Table 9.5 for details.)

on NHEL-CRP (Table 9.6; Plate 9.2). The ratio of the erosion rate on HEL:NHEL
category was 3.4 for 1987, 3.1 for 1992, and 3.1 for 1997.

The rate of decline of the extent and magnitude of erosion on CRP land over
the 15-year period from 1982 to 1997 is shown by regression equations presented
in Table 9.7 and Figure 9.4.

**Table 9.7   The Extent and Magnitude of Erosion on CRP Land**

| Dependent Variable | Regression Equation | $R^2$ |
|---|---|---|
| Extent of erosion on CRP land (Mha) | | |
| (i) Total CRP land | Y = 2.25 − 0.204x | 0.96 |
| (ii) HEL-CRP land | Y = 1.67 − 0.152x | 0.93 |
| (iii) NHEL-CRP land | Y = 0.50 − 0.053x | 0.87 |
| Magnitude of erosion on CRP land ($10^6$ Mg) | | |
| (i) Total CRP land | Y = 102 − 9.03x | 0.90 |
| (ii) HEL-CRP land | Y = 85.6 − 7.73x | 0.89 |
| (iii) Non-HEL-CRP land | Y = 16.2 − 1.3x | 0.95 |
| Rate of erosion on CRP land (Mg/ha/yr) | | |
| (i) Total CRP land | Y = 17.2 − 1.83x | 0.80 |
| (ii) HEL-CRP land | Y = 23.6 − 2.5x | 0.80 |
| (iii) NHEL-CRP land | Y = 6.9 − 0.72x | 0.81 |

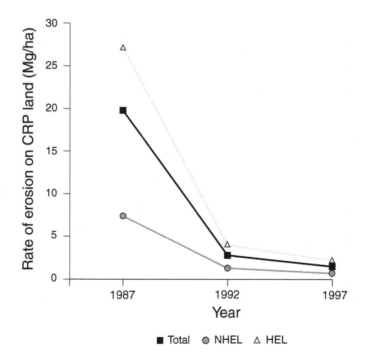

**Figure 9.4**   The rate of erosion on CRP land.

## 9.3 WIND EROSION IN THE U.S.

The data in Table 9.8 show the magnitude and rate of wind erosion in the U.S. The magnitude of soil displaced by wind erosion was 1.36 billion Mg in 1982 and declined to 0.77 billion Mg in 1997. The average rate of decline over the 15-year period was 66 million Mg/yr or 4.9%/yr. The average rate of wind erosion was 8.0 Mg/ha/yr in 1982 and 5.1 Mg/ha/yr in 1997, a decrease of 36% over the 15-year

Table 9.8   Wind Erosion in the U.S.

| Year | Wind Erosion (billion Mg) | Rate (Mg/ha/yr) |
|------|---------------------------|-----------------|
| 1982 | 1.36 | 8.00 |
| 1987 | 1.31 | 8.04 |
| 1992 | 0.94 | 6.05 |
| 1995 | 0.85 | 5.62 |
| 1996 | 0.77 | 5.44 |
| 1997 | 0.77 | 5.08 |

Source: Recalculated from Uri, N.D. and Lewis, J.A., Sci. Total Environ., 218, 45–48, 1998.

Table 9.9   Estimates of the Water, Wind, and Total Erosion in Conterminous United States

| Region | Area (Mha) | Magnitude of Erosion ($10^6$ Mg/yr) | | |
|--------|------------|-------|------|-------|
| | | Water | Wind | Total |
| NE Atlantic | 43.7 | 80 | 0 | 80 |
| SE Atlantic | 71.1 | 140 | 0 | 140 |
| Great Lakes | 45.6 | 70 | 140 | 210 |
| Mississippi Basin | 325.5 | 1510 | 1220 | 2730 |
| NW Gulf of Mexico | 81.4 | 190 | 880 | 1070 |
| Colorado Basin | 66.3 | 130 | 1570 | 1700 |
| Central Basin | 35.5 | 90 | 700 | 790 |
| NW Pacific | 71.4 | 190 | 150 | 340 |
| SW Pacific | 42.0 | 110 | 200 | 310 |
| Total | 782.5 | 2500 | 4860 | 7360 |

Source: Modified from Smith et al., Global Biogeochemical Cycles, 15, 697–707, 2001.

period. Similar to the water erosion, adoption of conservation-effective farming systems also reduced the risks of wind erosion. Conservation-effective systems were conversion of HEL to CRP, no till or conservation till, elimination of summer fallow, growing winter cover crops, establishment of conservation buffers and riparian filters, and controlled grazing.

## 9.4   TOTAL WIND AND WATER EROSION IN THE U.S.

Several attempts have been made to assess the extent of water and wind erosion in the U.S. Smith et al. (2001) used the NRI database to compute regional estimates of water and wind erosion on conterminous U.S. excluding the federal lands. The data in Table 9.9 show regional distribution of wind, water, and total erosion. Soil erosion was estimated at 2500 million Mg by water erosion, 4860 million Mg by wind erosion, and 7360 million Mg by total erosion.

The total soil moved by wind and water erosion in the U.S. was 3.05 billion Mg in 1982, 2.72 billion Mg in 1987, 2.02 billion Mg in 1992, 1.79 billion Mg in 1995, 1.75 billion Mg in 1996, and 1.78 billion Mg in 1997 (Figure 9.5). The magnitude of total soil erosion decreased at an average rate of 85 million Mg/yr. Uri (2000) reported total sheet, rill, and wind erosion in the U.S. from 1982 to 1992. Uri (2000)

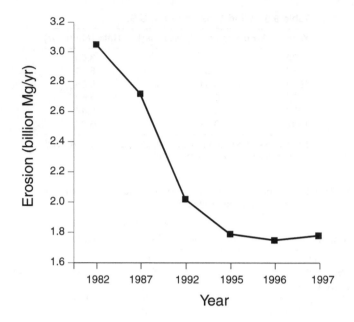

**Figure 9.5**   Total wind and water erosion in the U.S.

**Map 9.1**   Annual wind and water erosion (Uri, 2000). *Note*: One black dot equals 40,000,000 tons (36.3 million Mg) of soil loss due to water erosion based on the USLE. One gray dot equals 40,000,000 tons (36.3 million Mg) of soil loss due to wind erosion based on the WEQ. (From Uri, N.D., *J. Sust. Agric.*, 16, 71–94, 2000. With permission.)

prepared a map (Map 9.1) showing hot spots of wind and water erosion in 1992. West of the 100th meridian, the most significant sheet and rill erosion hot spots are in the Palouse region of Washington, Oregon, Idaho (Snake River Valley in eastern Idaho), and the eastern Colorado Plains. In contrast, wind erosion is primarily a concern in the southwest and western U.S. Accelerated soil erosion is also of a local concern in the Appalachian region of the Carolinas, Georgia, and Kentucky.

Total erosion on cropland and CRP land has been estimated at 2.8 billion Mg in 1982 and only 1.73 billion Mg in 1995 and 1997 (NRCS, 2000). These data indicated that since 1982, erosion on cropland and CRP land has been reduced by 38%. The NRI also indicates the following (NRCS, 2000):

- Since 1995, the erosion rate has not decreased and has leveled off to about 1.73 billion Mg/yr.
- About 45.3 Mha or 30% of the nation's cropland has been determined to be excessively eroding at the rate of 1.18 billion Mg/yr.
- Excessive erosion remains to be a problem on about 24.3 Mha of HEL cropland, and on 21.1 Mha of NHEL cropland.

Excessive erosion estimated at 1.18 billion Mg/yr can have severe adverse impacts on transport of sediments, nutrients, and pesticides in natural water. In addition to the adverse impact on air quality by wind erosion in some regions (West, Midwest, Northern Plains, and Southern Plains), excessive erosion also causes emission of GHGs into the atmosphere (Lal, 1995; Lal et al., 1998).

Most of the Farm Bills of the 1980s and 1990s targeted the HELs with an erodibility index of 8 or more (see Chapter 5). Uri (2000) compiled a map (Map 9.2) showing areas with high percentages of relatively high EI cropland. These lands were converted to forestland with perennial ground cover and low risks of soil erosion (Plate 9.3 to Plate 9.5).

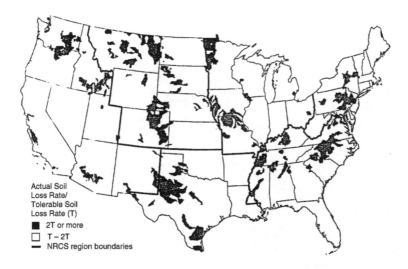

**Map 9.2**  Average annual soil erosion by wind and water on cultivated cropland as a proportion of the tolerable rate (T), 1992. (From Uri, N.D., *J. Sust. Agric.*, 16, 71–94, 2000. With permission.)

**Plate 9.3**  Establishment of trees on nonprime cropland decreases risks of soil erosion.

**Plate 9.4**  Trees provide a protective ground cover throughout the year.

**Plate 9.5**   In addition to controlling soil erosion, afforestation also leads to carbon sequestration in soil and biomass.

## 9.5 SEDIMENT LOAD IN U.S. RIVERS

Soil erosion data reported in Chapter 5 through Chapter 9 are based on estimates obtained through application of the Revised Universal Soil Loss Equation (RUSLE) and the Wind Erosion Equation as per methodology outlined in Chapter 4. These data estimate the magnitude of soil displaced by water (sheet and rill erosion) or wind. The delivery ratio, the fraction of sediment displaced that is eventually transported into the river system, reflects the amount of sediments transported out of the watershed. The data in Table 9.10 show the amount of sediments actually transported into the drainage system. The 503 Mha of the drainage/watershed area discharging 1124 km³ of water transports about 404 million Mg of sediments/yr.

## 9.6 CONCLUSIONS

Between 1982 and 1997, there was a drastic decline in both total erosion and erosion rate on U.S. cropland, pastureland, and minor lands. Further, the decline was most pronounced on HEL because it was specifically targeted by various Farm Bills of the 1980s and 1990s. As one would expect, total erosion (sheet, rill, and wind) declined during this period. Regression equations in Table 9.11 show that the magnitude of total erosion declined at the rate of $0.094 \times 10^9$ Mg/yr. The corresponding decline in the magnitude of erosion was $0.0432 \times 10^9$ Mg/yr for wind erosion and $0.047 \times 10^9$ Mg/yr for water (sheet and rill) erosion.

Table 9.10   Total Sediment Load in U.S. Rivers by Both Natural and
             Anthropogenic Processes

| River | Drainage Area (10⁶ km²) | Water Discharge (km³/yr) | Sediment Discharge (10⁶ Mg/yr) |
|---|---|---|---|
| Hudson | 0.02 | 12 | 1 |
| Mississippi | 3.27 | 580 | 210 |
| Brazos | 0.11 | 7 | 16 |
| Eel | 0.008 | — | 14 |
| Columbia | 0.67 | 251 | 8 |
| Yukon | 0.84 | 195 | 60 |
| Copper | 0.06 | 39 | 70 |
| Sustina | 0.5 | 40 | 25 |
| Total | 5.028 | 1124 | 404 |

Source: From Leeden et al., The Water Encyclopedia, 2nd ed., Lewis Publishers, Chelsea, MI, 1991.

Table 9.11   Temporal Changes in Total Sheet, Rill, and Wind
             Erosion in the U.S. between 1982 and 1997

| Dependent Variable | Regression Equation | $R^2$ |
|---|---|---|
| Total erosion ($10^9$ Mg) | $Y = 3.08 - 0.094x$ | 0.97 |
| Wind erosion ($10^9$ Mg) | $Y = 1.42 - 0.043x$ | 0.93 |
| Water erosion ($10^9$ Mg) | $Y = 1.65 - 0.047x$ | 0.95 |

$X$ = the period in years. For example, x is 0 for 1982 and 15 for 1997.

The Food Security Act of 1985 was the first major piece of legislation to explicitly tie eligibility to receive agricultural program payments to conservation performance. The Federal Agricultural Improvement and Reform Act (FAIR) of 1996 modifies the conservation compliance provisions by providing farmers with greater flexibility in implementing conservation plans. These programs have put land owner stewardship at an all-time high. Consequently, significant reductions in erosion have occurred because of the conversion of HEL to CRP and other restorative land uses. Because off-site effects of erosion are more drastic than on-site effects, several public policies have to be designed to influence farmers' choices of production practices and technologies. These may involve education, technical assistance, financial assistance, land retirement, regulation and taxes, and research and development (Uri, 2000; Napier et al., 1999). Soil erosion control in the U.S. is a success story with numerous on-site and off-site benefits.

## REFERENCES

Lal, R. Global soil erosion by water and carbon dynamics. In *Soils and Global Change*. R. Lal, J.M. Kimble, E. Levine, and B.A. Stewart (Eds.), CRC/Lewis Publishers, Boca Raton, FL, 1995, 131–142.

Lal, R., J.M. Kimble, R.F. Follett, and C.V. Vole. *The Potential of U.S. Cropland to Sequester Carbon and Mitigate the Greenhouse Effect.* CRC/Lewis Publishers, Boca Raton, FL, 1998, 128 pp.

Leeden, Van der F., F.L. Troise, and D.K. Todd. *The Water Encyclopedia*, 2nd ed. Lewis
    Publishers, Chelsea, MI, 1991, 808 pp.
Napier, T.L., S.M. Napier, and J. Turdon (Eds.). *Soil and Water Conservation Policies and
    Programs: Successes and Failures.* CRC Press, SWCS, Boca Raton, FL, 1999, 640 pp.
NRCS. *National Resource Inventory: Background and Highlights.* NRCS, Washington, D.C.,
    2000.
Smith, S.V., W.H. Renwick, R.W. Buddemeir, and C.J. Crossland. Budgets of soil erosion
    and deposition for sediments and sedimentary organic carbon across the conterminous
    United States. *Global Biogeochem. Cycles* 15, 697–707, 2001.
Uri, N.D. Agriculture and the environment: the problem of soil erosion. *J. Sust. Agric.* 16,
    71–94, 2000.
Uri, N.D. and J.A. Lewis. The dynamics of soil erosion in U.S. agriculture. *Sci. Total Environ.*
    218, 45–58, 1998.

# Other Forms of Degradation

# CHAPTER 10

# Desertification

Desertification is defined as "land degradation in arid, semiarid, and dry subhumid areas resulting mainly from adverse human impact" (UNEP, 1992). Desertification can also occur in cold and humid climates, especially under ecologically sensitive ecoregions (Arnalds et al., 2001). There are two important points in this definition. One, desertification addresses the problem of land rather than soil. The term land is broad-based and refers to an ecosystem that comprises climate, vegetation, soil, terrain, and hydrology. Two, desertification addresses the problem originating from adverse human impact or anthropogenic activities. Therefore, drought and other natural events merely exacerbate the problem but are not a driving factor in the process of desertification per se. The effects from the dust bowl of the 1930s were exacerbated by excessive tillage.

There are five principal processes of desertification: vegetation degradation, water erosion, wind erosion, salinization and waterlogging, and soil crusting and compaction (Dregne, 1998). In addition, soil pollution and contamination by land application of industrial waste, city sludge, pesticides, and other chemicals may also cause desertification. It is important that soil pollution can also occur in humid regions. Depletion of soil organic matter is also an important factor in soil degradation that leads to desertification.

The criteria used to assess the severity of desertification are based on the impact on long-term plant productivity. The range of loss in productivity for different severity classes of desertification or land degradation is given in Table 10.1. The assessment of desertification made by Dregne (1994) was based on "structured informed opinion analyses" (Dregne, 1989). This assessment method involves four steps: (1) collect all available information on soils, vegetation, geology, crop production, population distribution, and physiography of the region, (2) use a computer-assisted expert system that asks local knowledgeable persons (both professionals and nonprofessionals) a series of questions on specific degradation processes and their on-site and off-site impacts, (3) convene a meeting with a group of knowledgeable people to draw conclusions from step 2, and then bring different groups together to review the first conclusions and so on, and (4) finalize the conclusions and prepare

Table 10.1    Criteria to Assess the Severity of Land
              Degradation

| Land Use | % Loss of Potential Productivity | | | |
|---|---|---|---|---|
| | Slight | Moderate | Severe | Extreme |
| Irrigated land | 0–10 | 10–25 | 25–50 | 50–100 |
| Rainfed land | 0–10 | 10–25 | 25–50 | 50–100 |
| Rangeland | 0–25 | 25–50 | 50–75 | 75–100 |

*Source*: Modified from Dregne, H.E., *Soil Sci. Soc. Amer.*, Special publication 41, 53–58, 1994 and Dregne, H.E., in *Methods for Assessment of Soil Degradation*, CRC Press, Boca Raton, FL, 1998, 441–458.

Table 10.2    Dry Land Areas of the U.S.

| Land Use | Area (Mha) | % of Total Dry Land |
|---|---|---|
| Irrigated land | 15.2 | 4.1 |
| Rainfed cropland | 30.1 | 8.1 |
| Rangeland | 325.1 | 87.5 |
| Hyperarid | 1.3 | 0.3 |
| Total | 371.7 | 100 |

*Source*: Modified from Dregne, H.E. and Chou, N.-T., in *Degradation and Restoration of Arid Lands*, Texas Tech University, Lubbock,1992.

maps, which are reviewed by experienced individuals. What follows are the estimates of desertification in the U.S. based on the methods described above.

## 10.1 ARID LANDS IN THE U.S.

Arid lands in the conterminous U.S. are located in the regions where evapotranspiration exceeds the precipitation. Total dryland area in the U.S. is estimated at 372 Mha (Table 10.2). The total dry land area comprises 15.2 Ma (4.1%) of irrigated land, 30.1 Mha (8.1%) of rainfed cropland, 325.1 Mha (87.5%) of rangeland, and 1.3 Mha (0.3%) of hyperarid land. While "aridity" of these regions is a natural process, its effects are exacerbated and spread to adjacent regions by anthropogenic activities. Principal activities that exacerbate the severity and magnitude of desertification are overgrazing for prolonged period of time, removal of protective vegetal cover and change in biodiversity, especially the species mix leading to predominance of undesirable species, removal of natural vegetal cover, use of water on irrigated areas with high salts, and no leaching of the salts, etc.

## 10.2 THE EXTENT AND SEVERITY OF DESERTIFICATION

The extent (area) and severity (impact on biological productivity) for different land uses are shown in Table 10.3. The extent of desertification for moderate+ categories is estimated at 4.0 Mha (26.3%) for irrigated land, 3.6 Mha (12.0%) for

Table 10.3  Estimates of Desertification in the U.S.

| Land Use | Total Area | Slight | Moderate | Severe | Extreme | Total (Moderate+) | % Desertified |
|---|---|---|---|---|---|---|---|
| | | | Mha | | | | |
| Irrigated land | 15.2 | 11.2 | 3.5 | 0.4 | 0.1 | 4.0 | 23.6 |
| Rainfed cropland | 30.1 | 26.5 | 3.5 | 0.09 | 0.02 | 3.6 | 12.0 |
| Rangeland | 325.1 | 49.1 | 98.0 | 173.0 | 5.0 | 276.0 | 84.9 |

*Source*: Modified from Dregne.

rainfed cropland, and 276.0 Mha (84.9%) for rangeland. The desertification of rangeland comprises degradation of both vegetation and soil. Degradation of vegetation (e.g., reduction in ground cover, total biomass, and species diversity) leads to reduction in the SOC pool, a decline in overall soil quality, and a decrease in biomass productivity. Vegetation and soil degradation are mutually enhancing processes.

## 10.3 DEGRADATION OF RANGELAND VEGETATION

The degradation of rangeland vegetation is primarily caused by excessive prolonged grazing and removal of vegetation cover by anthropogenic perturbations. Excessive grazing affects the vegetation cover and species composition (Lal, 2001). Consequently, some rangelands are severely degraded. In the western U.S., only about 40% of the 160 Mha of rangeland is in good condition for forage production (USDA, 1987). About 70 to 90% of the ground cover is needed to effectively reduce risks of soil erosion. This much cover translates into 4 to 6 Mg/ha of herbage and litter for effective erosion control. There is an exponential decrease in sheet (interrill) erosion with increase in litter accumulation (Thurow et al., 1988). Soil erosion becomes a severe problem on rangelands if the spots of bare soil are larger than 10 cm in diameter in wheat grass type and 5 cm in the annual type of vegetation cover (Heady and Child, 1994). The effect of fire on rangeland is a complex issue to resolve. Most rangelands have coevolved with fire and grazing (Rice and Owensby, 2000). Therefore, use of fire as a management tool can be restorative to rangeland quality. It is the anthropogenic changes in species composition which, in interaction with intense burning, can cause adverse impacts on rangeland quality.

The extent of degraded rangeland vegetation in the U.S. is shown in Table 10.4. Total degraded rangeland vegetation (moderate+) in NRI 1992 assessment was 127.3 Mha or 76.3% of the total area. This comprised 124.5 Mha (77.2%) of the grazed and 2.8 Mha (50.9%) of the nongrazed rangeland. It is apparent that grazing is the principal cause of degradation of the rangeland vegetation.

Temporal changes on land area affected by vegetation degradation on rangeland between 1982 and 1992 are shown in Table 10.5. Total area affected by vegetation degradation was 2.72 Mha of nongrazed and 96.42 Mha of grazed rangeland in 1982, 3.80 Mha of nongrazed and 137.21 Mha of grazed rangeland in 1987, and 3.10 Mha of nongrazed and 108.12 Mha of grazed rangeland in 1992.

Table 10.4 Degradation of Rangeland Vegetation (1992 Assessment)

| Management | Total Area | Severity of Degradation | | | | Total (Moderate+) | % of Total Area |
| --- | --- | --- | --- | --- | --- | --- | --- |
| | | Slight | Moderate | Severe | Extreme | | |
| | | Mha | | | | | |
| Nongrazed | 5.54 | 0.95 | 1.72 | 0.61 | 0.49 | 2.82 | 50.9 |
| Grazed | 161.2 | 18.15 | 74.63 | 31.14 | 18.7 | 124.47 | 77.2 |
| Total | 166.76 | 19.10 | 76.35 | 31.75 | 19.19 | 127.29 | 76.3 |

Table 10.5 Vegetation Degradation on Rangeland

| Range Quality | 1982 | | 1987 | | 1992 | |
| --- | --- | --- | --- | --- | --- | --- |
| | Nongrazed | Grazed | Nongrazed | Grazed | Nongrazed | Grazed |
| | Mha | | | | | |
| Poor range condition | 0.92 | 24.92 | 1.09 | 24.21 | 0.84 | 22.47 |
| Range trend down | 0.44 | 23.29 | 0.81 | 28.45 | 0.71 | 32.61 |
| Poor or range trend down | 1.08 | 38.19 | 1.36 | 40.81 | 1.28 | 43.78 |
| Poor and range trend down | 0.28 | 10.02 | 0.54 | 11.85 | 0.27 | 9.26 |
| Total | 2.72 | 96.42 | 3.80 | 137.21 | 3.10 | 108.12 |

**Table 10.6    Area of Rangeland Degraded through Concentrated Flow (1992 Assessment)**

| Management | Total Area | Slight | Moderate | Severe | Extreme | Total (Moderate+) | % of Total Area |
|---|---|---|---|---|---|---|---|
| | | | Severity of Degradation | | | | |
| | | | Mha | | | | |
| Nongrazed | 5.54 | 1.92 | 0.005 | 0.03 | 0.0 | 0.04 | 0.7 |
| Grazed | 161.2 | 67.20 | 0.34 | 0.55 | 0.11 | 1.00 | 0.6 |
| Total | 166.76 | 69.12 | 0.35 | 0.58 | 0.11 | 1.04 | 0.6 |

**Table 10.7    Area of Rangeland Degraded through Wind Erosion (1992 Assessment)**

| Management | Total Area | Slight | Moderate | Severe | Very Severe | Total (Moderate+) | % of Total Area |
|---|---|---|---|---|---|---|---|
| | | | Severity of Degradation | | | | |
| | | | Mha | | | | |
| Nongrazed | 5.54 | 0 | 0.14 | 0.88 | 0.34 | 1.36 | 24.5 |
| Grazed | 161.2 | 0.02 | 1.83 | 25.22 | 16.53 | 43.58 | 27.0 |
| Total | 166.76 | 0.02 | 1.97 | 26.10 | 16.87 | 44.94 | 26.9 |

Therefore, land area affected by vegetation degradation progressively increased from 99.14 Mha in 1982 to 141.10 Mha in 1987. The degraded vegetation area was 111.22 Mha in 1992 (Table 10.5), a decrease from 1987.

## 10.4 RANGELAND DEGRADATION BY CONCENTRATED FLOW AND WIND EROSION

The effect of concentrated flow on rangeland degradation is shown by the data in Table 10.6. Total area affected by concentrated flow (moderate+) is merely 0.6% (1.0 Mha) of the total range land area of 166.8 Mha. In contrast, a large area of grazed rangeland is affected by wind erosion (Table 10.7). The area affected by wind erosion (moderate+) is 1.4 Mha (24.5%) for the nongrazed rangeland compared with 43.6 Mha (27.0%) for the grazed rangeland. The area of rangeland affected by wind erosion is 44.9 Mha or 26.9% of the total rangeland area.

## 10.5 DESERTIFICATION CONTROL

Desertification control involves reversal of the four processes of desertification, namely, restoring vegetation (controlling grazing), preventing or minimizing risks of water and wind erosion, restoring salinized soils by leaching salts out of the root zone through provision of drainage, and improving soil structure. In addition to enhancing biological productivity, desertification control also sequesters C in soil and vegetation, and reduces the rate of enrichment of atmospheric concentration of $CO_2$ (Lal, 2001).

The principal strategy is to restore the quality of rangeland vegetation and soil. There is a strong link between soil quality and vegetal cover and its quality. An improvement in soil quality enhances biomass production and vice versa. Resilience of rangeland is directly related to soil quality, and the extent and rate of rangeland recovery are improved by improvement in soil quality. Rangeland resilience depends on plant population dynamics, community interaction, species composition and biodiversity, and interaction of these vegetal parameters with soil properties. In addition, processes governing rangeland resilience depend on management (e.g., stocking rate, burning, nutrient and species management).

## REFERENCES

Arnalds, O., E.F. Borarinsdottir, S. Metusalemsson, A. Jonsson, and O.A. Arnason. *Soil Erosion in Iceland*. Soil Conservation Service, Agricultural Research Institute, Reykjavik, Iceland, 2001.

Dregne, H.E. Informed opinion: filling the soil erosion data gap. *J. Soil Water Cons.* 44, 303–305, 1989.

Dregne, H.E. Land degradation in the world's arid zones. In *Soil and Water Science: Key to Understanding Our Global Environment*. Soil Sci. Soc. Amer., Special Publ. 41, Madison, WI, 1994, pp. 53–58.

Dregne, H.E. Desertification assessment. In *Methods for Assessment of Soil Degradation*. R. Lal, W.H. Blum, C. Valentine, and B.A. Stewart (Eds.). CRC, Boca Raton, FL, 1998, pp. 441–458.

Dregne, H.E. and N-T. Chou. Global desertification dimensions and costs. In *Degradation and Restoration of Arid Lands*. H.E. Dregne (Ed.). ICASALS, Texas Tech University, Lubbock, 1992, pp. 249–281.

Heady, H.F. and R.D. Child. *Rangeland Ecology and Management*. Westview Press, Boulder, CO, 1994.

Lal, R. Potential of desertification control to sequester carbon and mitigate the greenhouse effect. *Climate Change* 15, 35–72, 2001.

Lal, R. Soil erosion and carbon dynamics on grazing land. In *The Potential of U.S. Grazing Lands to Sequester Carbon and Mitigate the Greenhouse Effect*. R.F. Follett, J.M. Kimble, and R. Lal (Eds.), CRC, Boca Raton, FL, 2001, pp. 231–247.

Rice, C.W. and C.E. Owensby. The effects of fire and grazing on soil carbon in rangelands. In *The Potential of U.S. Grazing Lands to Sequester Carbon and Mitigate the Greenhouse Effect*, R.F. Follett, J.M. Kimble, and R. Lal (Eds.), CRC/Lewis Publishers, Boca Raton, FL, 2001, pp. 323–342.

Thurow, T.L., W.H. Blackburn, and C.A. Taylor, Jr. Infiltration and inter-rill erosion responses to selected livestock grazing strategies, Edwards Plateau Texas. *J. Range Manage.* 41, 296–302, 1988.

UNEP. *World Atlas of Desertification*. Edward Arnold, Hodder & Stoughton, Sevenoakes, U.K., 1992, 69 pp.

USDA. The Second RCA Appraisal. Soil, Water and Related Resources on NonFederal Lands in the United States. USDA, Washington, D.C., 1987.

# Chapter 11

# Salinization

Excessive accumulation of salt in the root zone occurs in soils of arid and semiarid regions with high evaporation losses (Plate 11.1). These salt-affected soils are formed under climatic conditions with average rainfall below 500 mm/yr and high evaporation. In such conditions, deep percolation is limited resulting in restricted leaching of accumulated salts.

There are three types of salt-affected soils: saline soils, saline-sodic soils, and alkali soils. Saline soils are recognized by the presence of white salt encrustations on the surface. These soils contain high concentrations of chlorides, sulfates of Na, Ca, and Mg. The electrical conductivity of saturation extracts of saline soils is generally more than 4 dS/m at 25C. The sodium absorption ratio (SAR = Na/[(Ca + Mg)/2]$^{1/2}$) of the soil solution is usually less than 15. Salt-affected soils with SAR >15 are usually termed saline-sodic soils. In comparison, saline-sodic soils have neutral soluble salts of Na. Leaching of saline-sodic soils with good quality water and the addition of gypsum may result in desalinization and desodification. The alkali soils contain an excess of exchangeable sodium and contain sodium carbonates. Most soils in the arid and semiarid regions also contain $CaCO_3$. Such calcareous soils contain soluble carbonates, have high exchangeable Na percentage, usually >15, and pH of the saturated paste >8.2. The electrical conductivity of the saturated extracts is usually more than 4 dS m$^{-1}$ at 25C (Gupta and Abrol, 1990; Szabolcs, 1998).

Sodic soils are also subject to severe structural degradation. High concentrations of Na$^+$ on the exchange complex disperse clay, and swelling of sodic aggregates destroys soil structure. Dispersion of clay and breakdown of aggregates reduce the porosity and permeability of soils when wet, and increase strength when dry. Sodic soils have adverse soil physical conditions with a limited range of water availability (Rengasamy, 1998; de Silva et al., 1994).

There are two factors leading to formation of salt-affected soils: (1) source of soluble salts, and (2) restricted leaching because of lack of water and/or presence of impermeable subsoil with impeded drainage. The nature of salt accumulation depends on: (1) the quantity of water soluble salts, (2) the chemistry of salinization, and (3) the distribution of salts in the soil profile (Szabolcs, 1998).

**Plate 11.1**   Salt-affected soils occur in arid and semiarid regions.

There are two types of salinization processes: (1) primary or natural, and (2) secondary or anthropogenic. Most salt-affected soils are formed due to natural processes related to geology, hydrology, terrain, and soil characteristics. The secondary salinization, however, occurs due to human activities. The principal cause of secondary salinization is excessive irrigation, poor quality water, and lack of adequate drainage. There is a strong correlation between the increase in the area affected by secondary salinization and the area under irrigation (Szabolcs, 1998). In addition to irrigation, other anthropogenic processes that exacerbate salinization include deforestation, overgrazing, change in land use and cultivation patterns, depletion of freshwater layers in the vicinity of the soil surface, fallowing land, changing or removing natural drainage patterns, and chemical contamination.

## 11.1  IRRIGATED LAND AREA IN THE U.S.

The extent of irrigated land area changes with time depending on the competition for water and economic factors. Estimates of the irrigated land area in the U.S. are shown in Table 11.1 and Table 11.2. These data are relevant because the problem of secondary salinization is the most pronounced on lands irrigated with poor quality water and with slow or impeded drainage. The data in Table 11.1, based on estimates reported by FAO, show that irrigated land area in the U.S. was less than 17 Mha in the 1970s, less than 20 Mha in the 1980s, and about 21.4 Mha in the 1990s. The data in Table 11.2, based on the NRCS survey (NRCS, 2000), show total (cultivated and noncultivated) irrigated cropland area was 24.7 Mha in 1982, 24.8 Mha in 1987, and 25.2 Mha in 1992.

**Table 11.1 Irrigation in the U.S.**

| Year | Land Area (Mha) |
|------|-----------------|
| 1977 | 16.7 |
| 1982 | 19.8 |
| 1987 | 18.8 |
| 1990 | 20.8 |
| 1992 | 20.3 |
| 1993 | 21.4 |
| 1994 | 21.4 |
| 1995 | 21.4 |
| 1996 | 21.4 |
| 1997 | 21.4 |

*Source:* Modified from FAO, *Production Yearbook*, vols. 47, 50, 52, Rome, 1993, 1996, 1998.

**Table 11.2 Irrigated Agricultural Land in the U.S.**

| Land Use | Year | Total Area (Mha) |
|----------|------|-------------------|
| Cropland | | |
| (i) Cultivated | 1982 | 19.5 |
| | 1987 | 19.2 |
| | 1992 | 19.3 |
| (ii) Noncultivated | 1982 | 5.5 |
| | 1987 | 5.6 |
| | 1992 | 5.9 |

*Source*: NRI, Washington, D.C., 1992.

**Table 11.3 Changes in Irrigated Land Area in the U.S. from 1982 to 1997**

| Region | Change in Irrigated Land Area | | | |
|--------|-------------|-------------|-------------|-------------|
| | 1982–1987 | 1987–1992 | 1992–1997 | 1982–1997 |
| | ha | | | |
| Northeast | 3198.4 | 20202.4 | 19514.2 | 42915.0 |
| Lake States | 14493.9 | 48664.0 | −25384.6 | 37773.3 |
| Corn Belt | −6234.8 | 113400.8 | 17489.9 | 124655.9 |
| Northern Plains | 123724.7 | 131740.9 | 83805.7 | 339271.3 |
| Appalachia | 24413.0 | 4939.3 | 24534.4 | 53886.6 |
| Southeast | −19595.1 | 52145.7 | −64534.4 | −31983.8 |
| Delta States | 103562.8 | 418583.0 | 42429.1 | 564574.9 |
| Southern Plains | 291659.9 | −199149.8 | −172429.1 | −663238.9 |
| Mountain | −65749.0 | −100890.7 | −107085.0 | −273724.7 |
| Pacific | 6842.1 | −204858.3 | −160769.2 | −358785.4 |
| Total | −107004.0 | 284777.3 | −342429.1 | −164655.9 |

*Source*: Recalculated from NCRS, *National Resources Inventory: Background and Highlights*, Washington, D.C., 2000.

Changes in irrigated land area in the U.S. between 1982 and 1997 for a 5-year period are shown in Table 11.3 (NRCS, 2000). The greatest shift in distribution of irrigated land area occurred from 1987 to 1992. This trend continued from 1992 to 1997, but at a lower rate. These data show the following:

- The irrigated land area continued to decline in the western and south-central states but increased in the Midwest, Northern Plains, East, and Southeast.
- The largest decrease in irrigation since 1982 occurred in Texas, California, Nevada, and Idaho.
- On a regional basis, the largest decline in irrigation occurred in the Southern Plains, Pacific, and Mountain states. In comparison, the largest increase occurred in the Delta, Northern Plains, Corn Belt, Appalachian, Northeast, and Great Lakes states.
- The largest increases in irrigated area since 1982 occurred in Arkansas, Nebraska, Mississippi, Missouri, Michigan, and North Carolina.

## 11.2 CROPLAND AFFECTED BY SALINIZATION

The cropland area reportedly affected by salinization from the NRI in 1987, 1992, and 1997 is shown in Table 11.4. Total area affected by salinization was 0.67 Mha in 1987, 0.28 Mha in 1992 and 6.39 Mha in 1997. Considering a moderate+ level of salinization, the cropland area affected by salinization was 0.47 Mha in 1987, 0.23 Mha in 1992, and 1.97 Mha in 1997 (Table 11.4). Because irrigation is a major factor in salinization, the NRI assessments in 1987 and 1992 were made separately for irrigated cropland. Total area under irrigated cropland (cultivated and noncultivated) was 24.8 Mha in 1987 and 25.2 Mha in 1992. Irrigated cropland affected by salinization was 0.39 Mha in 1987 and 0.19 Mha in 1992. The drastic increase in salinization of cropland in 1997 may in part be due to differences in criteria used for assessment of soil degradation.

## 11.3 GRAZING LAND AFFECTED BY SALINIZATION

Grazing land comprises pastureland, rangeland, and forestland. The data in Table 11.5 show that salinization is not a problem in rangeland and forestland. The data on pastureland affected by salinization show that soils in grazed pastureland are more prone to salinization than those in ungrazed pasture or the idle land. The area affected by salinization in 1987 was 4.5 Mha (50.6%) in nongrazed pasture and 22.1 Mha (51.8%) in grazed pasture. Similar trends were observed in 1992. The area affected by salinization was 1.8 Mha (34.6%) in nongrazed pastures and 11.8 Mha (25.8%) in grazed pastures. The 1997 assessment did not separate the pastureland into nongrazed and grazed areas. The pastureland area affected by salinization was only 0.33 Mha or 0.7% of the total pasture area.

## 11.4 OTHER LANDS AFFECTED BY SALINIZATION

The data in Table 11.6 show minor land affected by salinization. The minor land area affected by salinization was 0.29 Mha in 1987, 0.37 Mha in 1992, and 0 Mha in 1997. Assessment in 1997 showed that 0.06 Mha of Conservation Reserve Program (CRP) land was affected by salinization.

Table 11.4 Cropland Area Affected by Salinization

| Land Use | Total Area | Light | Moderate | Severe | Extreme | Total | % of Total Area |
|---|---|---|---|---|---|---|---|
| | Mha | | | 10³ ha | | | |
| 1987 Assessment | | | | | | | |
| Cultivated cropland, nonirrigated | 122.9 | 194.5 | 19.6 | 20.9 | 45.1 | 280.1 | 0.23 |
| Cultivated cropland, irrigated | 19.2 | 0 | 21.5 | 79.4 | 234.5 | 335.4 | 1.7 |
| Noncultivated cropland, nonirrigated | 17.0 | 4.9 | 0.6 | 0 | 1.3 | 6.8 | 0.04 |
| Noncultivated cropland, irrigated | 5.6 | 0 | 6.5 | 7.3 | 37.0 | 50.8 | 0.91 |
| Total | 164.7 | 199.4 | 48.2 | 107.6 | 317.9 | 673.1 | |
| 1992 Assessment | | | | | | | |
| Cultivated cropland, nonirrigated | 112.5 | 49.4 | 8.7 | 11.9 | 15.0 | 85.0 | — |
| Cultivated cropland, irrigated | 19.3 | 0 | 2.6 | 25.7 | 136.5 | 164.8 | — |
| Noncultivated cropland, nonirrigated | 17.1 | 1.2 | 0 | 0 | 1.3 | 2.5 | — |
| Noncultivated cropland, irrigated | 5.9 | 0 | 4.3 | 2.6 | 19.3 | 26.2 | — |
| Total | 154.8 | 50.6 | 15.6 | 40.2 | 172.1 | 278.5 | — |

Table 11.5  Grazing Land Affected by Salinization on U.S. Private Land

| Land Use | Total Area | Light | Moderate | Severe | Extreme | Total | % of Total Area |
|---|---|---|---|---|---|---|---|
| | Mha | | | 1000 ha | | | |
| **1987 Assessment** | | | | | | | |
| Pastureland, nongrazed | 8.9 | 0 | 0.2 | 0.2 | 4.1 | 4.5 | 50.6 |
| Pastureland, grazed | 42.7 | 0 | 6.4 | 2.9 | 12.8 | 22.1 | 51.8 |
| Rangeland, nongrazed | 7.6 | 0 | 0 | 0 | 0 | 0 | 0 |
| Rangeland, grazed | 155.4 | 0 | 0 | 0 | 0 | 0 | 0 |
| Forestland, grazed | 25.7 | 0 | 0 | 0 | 0 | 0 | 0 |
| Forestland, nongrazed | 134.5 | 0 | 0 | 0 | 0 | 0 | 0 |
| **1992 Assessment** | | | | | | | |
| Pastureland, nongrazed | 5.2 | 0.0 | 0.0 | 0.0 | 1.8 | 1.8 | 34.6 |
| Pastureland, grazed | 45.8 | 0.0 | 4.5 | 1.2 | 6.1 | 11.8 | 25.8 |
| Rangeland, nongrazed | 5.2 | 0.0 | 0.0 | 0.0 | 0.0 | 0.0 | 0 |
| Rangeland, grazed | 156.4 | 0.0 | 0.0 | 0.0 | 0.0 | 0.0 | 0 |
| Forestland, grazed | 25.5 | 0.0 | 0.0 | 0.0 | 0.0 | 0.0 | 0 |
| Forestland, nongrazed | 134.4 | 0.0 | 0.0 | 0.0 | 0.0 | 0.0 | 0 |

**Table 11.6 Other Lands Affected by Salinization**

| Land Use | Total Area | Light | Moderate | Severe | Extreme | Total |
|---|---|---|---|---|---|---|
| | Mha | | | 1000 ha | | |
| 1987 Assessment | | | | | | |
| Minor land | 21.6 | 0 | 1.3 | 171.5 | 114.4 | 287.2 |
| Urban built-up land | 25.5 | 0 | 0 | 0 | 0 | 0 |
| 1992 Assessment | | | | | | |
| Minor land | 22.1 | 0.0 | 3.0 | 167.0 | 197.1 | 367.1 |
| Urban built-up land | 26.5 | 0.0 | 0.0 | 0.0 | 0.0 | 0.0 |
| 1997 Assessment | | | | | | |
| Minor land | 20.71 | 0.0 | 0.0 | 0.0 | 0.0 | 0.0 |
| CRP | 13.24 | 0.0 | 0.02 | 0.02 | 0.02 | 0.06 |

## 11.5 CONCLUSIONS

Croplands are most affected by salinization, followed by pastureland. In cropland, irrigated land is prone to secondary salinization. Land use, adoption of best management practices (BMPs), and improving soil drainage are important considerations for reducing risks of salinization. In addition to growing salt-tolerant crops, there are tillage methods that improve crop growth on salt-affected soils. Ridge tillage and growing crops in the furrow or half-way to the top of the ridge are good strategies.

## REFERENCES

da Silva, A.P., B.D. Kay, and E. Perfect. Characterization of the least limiting water range of soils. *Soil Sci. Soc. Am. J.* 58, 1775–1781, 1994.

FAO. *Production Yearbook*, Vol. 47. FAO, Rome, Italy, 1993.

FAO. *Production Yearbook*, Vol. 50. FAO, Rome, Italy, 1996.

FAO. *Production Yearbook*, Vol. 52. FAO, Rome, Italy, 1998.

Gupta, R.K. and I.P. Abrol. Salt-affected soils: their reclamation and management for crop production. In *Soil Degradation*. R. Lal and B.A. Stewart (Eds.), Advances in Soil Science 11, 223–288, 1990.

NRCS. National Resources Inventory: Background and Highlights. Washington, D.C., 2000.

Rengasamy, P. Sodic soils. In *Methods for Assessment of Soil Degradation*. R. Lal, W.H. Blum, C. Valentine, and B.A. Stewart (Eds.), CRC Press, Boca Raton, FL, 1998, 265–277.

Szabolcs, I. Salt build up as a factor of soil degradation. In *Methods for Assessment of Soil Degradation*. R. Lal, W.H. Blum, C. Valentine, and B.A. Stewart (Eds.), CRC Press, Boca Raton, FL, 1998, 253–264.

# CHAPTER 12

# Soil Degradation by Mining and Other Disturbance

Surface mining of coal and other minerals has been practiced for more than 150 years in the U.S. At the beginning of the 20th century, there were 14 surface coal mines near the small east Texas towns of Hoyt and Alba (Chapman, 1983). Surface mining was the common technique wherever the coal or other minerals were close to the surface. Soil disturbance and degradation due to mining bring about drastic changes in the original soil profile, and its properties and processes. Despite being an economical technique, however, surface mining leads to severe environmental concerns, especially with regard to acid mine drainage, severe erosion and sedimentation, and water pollution and contamination. Surface mining involves removal of the overburden from large continuous areas. Overburden is that part of the lithosphere that has to be moved before the mineral of interest is accessible. Surface mining, therefore, generates a voluminous amount of "spoil." It is a term given to any earth material, excluding the material being mined, that is left unmanaged. Minesoil is a mixture of soil and spoil or overburden that is being managed or reclaimed. Thus, reclaimed minesoils are manmade or anthropic soils.

Surface mining in the mountainous regions poses more severe environmental challenges. Spoil material deposited on the steep slopes can trigger landslides and mass movement. The problem of erosion and sedimentation is severe in mining sites in the mountainous terrain. Addressing the sedimentation problem necessitates restoration of mineral soils, and reestablishing the vegetal cover of grasses and/or trees. In addition, establishment of sediment traps and other filters is necessary to reduce risks of pollution and contamination of natural waters.

## 12.1 HISTORIC RECORDS OF SOIL DISTURBANCE BY MINING

Mining causes the most drastic soil disturbance (Plate 12.1 and Plate 12.2). Initially, coal was mined by the "pillar and room" method. This caused the problem of sink holes. Later, surface mining was done by using a big power shovel that digs

**Plate 12.1**   Exploration for ores, oil, and gas and other mining activities cause drastic soil disturbance.

**Plate 12.2**   Mining causes drastic disturbance of the soil.

into the soil, piling up the excavated material to expose the coal veins and other minerals. Similar magnitudes of disturbance are caused by a big auger (>2 m wide). By the mid-1970s, the land area disturbed by mining was estimated at about 0.25% of the land area of the U.S. (Chapman, 1983). As of July 1, 1977, the mining industry had disturbed 2.3 Mha, an area about the size of Vermont (Map 12.1). About 1.2 Mha of this land had been reclaimed by 1983. In the humid region alone, the land area

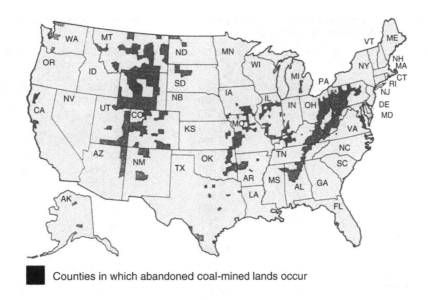

Counties in which abandoned coal-mined lands occur

**Map 12.1**  Land distributed by surface mining as of 1977 (From Chapman, E.W., in *Using our Natural Resources, 1983 Yearbook of Agriculture*, Washington, D.C., 394–403.)

affected by mining as of July 1, 1977 included 0.63 Mha by coal mines, 0.36 Mha by sand and gravel mining, and 0.30 Mha by other mining operations (Bastian et al., 1982) (Table 12.1). The total land area that needed reclamation, including dry regions, was 1.3 Mha in 1977 (Sutton and Dick, 1987).

## 12.2  SOIL DEGRADATION BY MINING

The data in Table 12.2 show the extent of soil degradation by mining. Soil erosion can be a severe hazard on mine soil (Plate 12.3). The soils disturbed by mining are used for different purposes including cropland, grazing land (pasture and range), forestland, and minor lands. There were no definite trends in soil degradation by mining on pastureland, except that the area declined during the 1990s. Most of the new mining disturbance occurs on rangeland. The rangeland area affected by mining decreased from 0.94 Mha in 1982 to 0.89 Mha in 1997, at an average rate of only 3.3 Tha/yr. There is quite a lot of mining activity on forestlands (Table 12.2), with the extent of soil disturbance ranging from 0.27 Mha in 1982 to 0.18 Mha in 1997. Over the 15-year period, soil disturbance by mining on forestland decreased at an average rate of 6.4 Tha/yr.

Total soil area disturbed by mining was 1.32 Mha in 1982, 1.29 Mha in 1987, 1.25 Mha in 1992, and 1.14 Mha in 1997. The average rate of decline of soil disturbance has been 11.9 Tha/yr over the 15-year period. The data in Table 12.3 show that area with extreme disturbance due to mining ranged between 1.83 Mha and 1.86 Mha over the 10-year period between 1982 and 1992. The severity of soil disturbance by mining is shown in Table 12.4. Considering all disturbance classes,

Table 12.1    Land Surface Mined in the Humid Region of the United States Needing
Reclamation as of July 1, 1977

| States | Coal Mines | Sand and Gravel | Other Mined Areas | Total |
|--------|-----------:|----------------:|------------------:|------:|
| | | ha | | |
| Alabama | 43,375 | 8,954 | 10,603 | 62,932 |
| Arkansas | 3,435 | 8,709 | 5,294 | 17,438 |
| Connecticut | — | 6,780 | 319 | 7,099 |
| Delaware | — | 1,179 | 26 | 1,205 |
| Florida | — | 5,884 | 103,931 | 109,815 |
| Georgia | 989 | 3,230 | 15,301 | 19,520 |
| Illinois | 64,642 | 11,710 | 7,594 | 83,946 |
| Indiana | 40,687 | 6,500 | 3,408 | 50,595 |
| Iowa | 5,807 | 7,535 | 6,504 | 19,846 |
| Kansas | 17,039 | 5,988 | 5,726 | 28,753 |
| Kentucky | 103,621 | 1,328 | 3,034 | 107,983 |
| Louisiana | — | 15,116 | 1,032 | 16,148 |
| Maine | — | 12,606 | 1,214 | 13,820 |
| Maryland | 4,907 | 6,954 | 1,180 | 13,041 |
| Massachusetts | — | 12,977 | 4,184 | 17,161 |
| Michigan | 58 | 22,310 | 11,135 | 33,503 |
| Minnesota | — | 17,209 | 21,340 | 38,549 |
| Mississippi | — | 18,616 | 3,168 | 21,784 |
| Missouri | 32,182 | 2,236 | 13,868 | 48,286 |
| Nebraska | — | 7,277 | 1,632 | 8,909 |
| New Hampshire | — | 5,154 | 169 | 5,323 |
| New Jersey | — | 9,967 | 2,256 | 12,223 |
| New York | — | 18,992 | 9,837 | 28,829 |
| North Carolina | — | 7,697 | 3,524 | 11,221 |
| Ohio | 110,872 | 15,909 | 11,077 | 137,858 |
| Oklahoma | 17,179 | 3,817 | 7,378 | 28,374 |
| Pennsylvania | 121,500 | 10,530 | 18,427 | 150,457 |
| Rhode Island | — | 1,050 | — | 1,050 |
| South Carolina | — | 5,451 | 2,156 | 7,607 |
| Tennessee | 13,247 | 2,333 | 1,394 | 16,974 |
| Texas | 2,850 | 64,292 | 17,048 | 84,190 |
| Vermont | — | 1,723 | 866 | 2,589 |
| Virginia | 12,938 | 3,125 | 1,318 | 17,381 |
| West Virginia | 37,473 | 1,844 | 403 | 39,720 |
| Wisconsin | — | 21,664 | 4,220 | 25,884 |
| Total | 632,801 | 356,646 | 300,566 | 1,290,013 |

*Source*: Bastian et al., in *Land Reclamation and Biomass Production with Municipal Wastewater and Sludge*, Penn State University Press, University Park, PA, 1982, 13–54; Sutton, P. and Dick, W.A., *Adv. Agron.* 41, 377–405, 1987.

total area disturbed by new land being mined was 2.19 Mha in 1982, 2.14 Mha in 1987, and 2.1 Mha in 1992.

The extent of cropland area disturbed by mining decreased over the 15-year period, at an average rate of 2.1 Tha/yr or 3.8%/yr (Figure 12.1). There was a similar decline in the extent of area degraded by mining in other land use categories between 1982 and 1997. The rate of decline in degraded land area was 2.1 Tha/yr for cropland, 0.066 Tha/yr for pastureland, 3.17 Tha/yr for rangeland, and 6.15 Tha/yr for forest-land (Table 12.5).

**Table 12.2   Soil Degradation by Mining or Other Soil Removal/Destruction**

| Land Use | 1982 | 1987 | 1992 | 1997 |
|---|---|---|---|---|
| | 10³ Ha | | | |
| Cultivated cropland | 47.57 | 38.46 | 31.78 | 22.71 |
| Noncultivated cropland | 9.59 | 11.22 | 9.39 | 2.33 |
| Pastureland | 53.27 | 50.28 | 56.88 | 50.00 |
| Rangeland | 939.4 | 935.00 | 920.10 | 891.60 |
| Forestland | 271.0 | 251.80 | 232.20 | 175.10 |
| Minor land | 0 | 0 | 0 | 0 |
| CRP | 0 | 0 | 0 | 0 |
| Total | 1320.83 | 1286.76 | 1250.35 | 1141.64 |

**Table 12.3   Extreme Soil Degradation by Mining or Other Soil Removal/Destruction on Private U.S. Minor Land (1992 Assessment)**

| Year | Total Area | Extreme Disturbance |
|---|---|---|
| | Mha | |
| 1982 | 21.39 | 1.83 |
| 1987 | 21.58 | 1.86 |
| 1992 | 22.10 | 1.86 |

**Plate 12.3**   Unreclaimed mine soil is prone to severe erosion with adverse impacts on water quality.

## 12.3 SOIL DISTURBANCE BY MINING ON CROPLAND

The data in Table 12.6 are based on the 1992 assessment, and show the severity of soil degradation in nonirrigated and irrigated cropland. There was no mining activity in irrigated cropland. The extent of disturbance in total cropland was 96.7 Tha in

TABLE 12.4   Total Land Area Affected by Mining or Other Soil
             Removal/Destruction on Private U.S. Land (1992
             Assessment)

| Year | Light | Moderate | Severe | Extreme | Total (Moderate +) |
|------|-------|----------|--------|---------|--------------------|
| | | | Area Affected by Mining | | |
| | | | Mha | | |
| 1982 | 0.0 | 0.082 | 0.290 | 1.822 | 2.194 |
| 1987 | 0.0 | 0.079 | 0.199 | 1.863 | 2.141 |
| 1992 | 0.0 | 0.182 | 0.059 | 1.857 | 2.098 |

Table 12.5   Regression Equations Relating Soil Degradation by Mining
             with Time between 1982 (Base Year) and 1997 (15 Years)

| Dependent Variable | Regression Equation | $R^2$ |
|--------------------|---------------------|-------|
| Cropland degraded by mining ($10^3$ ha) | $Y = 59.0 - 2.1x$ | 0.963 |
| Pastureland degraded by mining ($10^3$ ha) | $Y = 53.1 - 0.066x$ | 0.018 |
| Rangeland degraded by mining ($10^3$ ha) | $Y = 945.0 - 3.17x$ | 0.0896 |
| Forestland degraded by mining ($10^3$ ha) | $Y = 279.0 - 6.15x$ | 0.92 |

X = number of years. For example, x is 0 for 1982, 5 for 1987, 10 for 1992,
and 15 for 1997.

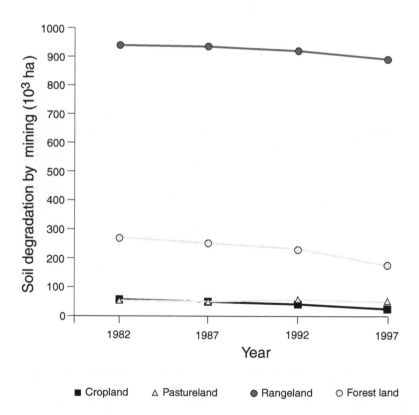

Figure 12.1   The extent of soil degradation by mining showing some reduction in the areas
             affected between 1982 and 1997.

**Table 12.6   Soil Degradation by Mining or Other Soil Removal/Destruction on Private U.S. Irrigated and Nonirrigated Cropland (1992 Assessment)**

| Land Use | Type | Year | Total Area | Area Affected by Mining | | | |
|---|---|---|---|---|---|---|---|
| | | | | Moderate | Severe | Extreme | Total |
| | | | Mha | 10³ ha | | | |
| Cultivated | Nonirrigated | 1982 | 128.76 | 7.45 | 54.58 | 0.0 | 62.03 |
| | | 1987 | 122.87 | 9.19 | 22.31 | 0.0 | 31.50 |
| | | 1992 | 112.46 | 3.00 | 0.0 | 0.0 | 3.00 |
| | Irrigated | 1982 | 19.50 | 0.0 | 0.0 | 0.0 | 0.0 |
| | | 1987 | 19.20 | 0.0 | 0.0 | 0.0 | 0.0 |
| | | 1992 | 19.31 | 0.0 | 0.0 | 0.0 | 0.0 |
| Noncultivated | Nonirrigated | 1982 | 16.67 | 15.87 | 18.79 | 0.0 | 34.66 |
| | | 1987 | 16.97 | 24.13 | 16.11 | 0.0 | 40.24 |
| | | 1992 | 17.15 | 30.12 | 4.94 | 0.0 | 35.06 |
| | Irrigated | 1982 | 5.50 | 0.0 | 0.0 | 0.0 | 0.0 |
| | | 1987 | 5.60 | 0.0 | 0.0 | 0.0 | 0.0 |
| | | 1992 | 5.87 | 0.0 | 0.0 | 0.0 | 0.0 |

1982, 71.7 Tha in 1987, and 38.1 Tha in 1992. There was no extreme disturbance, but severe disturbance on nonirrigated cropland occurred on 73.4 Tha in 1982, 38.4 Tha in 1987, and 4.9 Tha in 1992. Although the data show rather limited disturbance of cropland by mining activities, there are severe off-site problems that cannot be ignored. Some of the off-site problems include erosion, water contamination, sedimentation, and movement of sediment into cropland areas.

## 12.4  SOIL DISTURBANCE BY MINING ON GRAZING LAND

The data in Table 12.7 are based on 1992 assessment of soil disturbance on grazed and nongrazed pastureland and rangeland. Similar to irrigated cropland, there was neither any mining activity on grazed pasture nor on grazed rangeland. As one would expect, soil disturbance by mining was confined to nongrazed pastureland and nongrazed rangeland. The land area affected by mining disturbance on nongrazed pastureland was 141.9 Tha in 1982, 96.5 Tha in 1987, and 167.1 Tha in 1992. Similarly, the land area affected by mining disturbance on nongrazed pasture was 134.2 Tha in 1982, 108.8 Tha in 1987, and 36.0 Tha in 1992 (Table 12.7).

## 12.5  TOTAL SOIL DISTURBANCE BY MINING

The data in Table 12.8, based on 1992 NRI data, show total cumulative soil disturbance by all forms of mining in all states within the conterminous U.S. It is apparent that all land disturbed by mining in the 1992 NRI assessment was 4.4 Mha. In accordance with land distribution shown in Figure 12.1, mining-induced soil disturbance was relatively high in Pennsylvania, Texas, Florida, Kentucky, Minnesota, Wyoming, Michigan, Missouri, and Ohio.

Table 12.7 Soil Degradation by Mining or Other Soil Removal Destruction on Private U.S. Grazing Land Comprising Grazed and Nongrazed Pastures (1992 Assessment)

| Land Use | Type | Year | Total Area | Area Affected by Mining | | | |
|---|---|---|---|---|---|---|---|
| | | | | Moderate | Severe | Extreme | Total |
| | | | Mha | 10³ ha | | | |
| Pastureland | Grazed | 1982 | 40.16 | 0.0 | 0.0 | 0.0 | 0.0 |
| | | 1987 | 42.73 | 0.0 | 0.0 | 0.0 | 0.0 |
| | | 1992 | 45.78 | 0.0 | 0.0 | 0.0 | 0.0 |
| | Nongrazed | 1982 | 13.23 | 58.79 | 83.12 | 0.0 | 141.91 |
| | | 1987 | 8.93 | 45.26 | 51.26 | 0.0 | 96.52 |
| | | 1992 | 5.20 | 148.95 | 18.18 | 0.0 | 167.13 |
| Rangeland | Grazed | 1982 | 155.04 | 0.0 | 0.0 | 0.0 | 0.0 |
| | | 1987 | 155.43 | 0.0 | 0.0 | 0.0 | 0.0 |
| | | 1992 | 156.36 | 0.0 | 0.0 | 0.0 | 0.0 |
| | Nongrazed | 1982 | 10.51 | 0.0 | 134.17 | 0.0 | 134.17 |
| | | 1987 | 7.65 | 0.0 | 108.83 | 0.0 | 108.83 |
| | | 1992 | 5.16 | 0.0 | 36.03 | 0.0 | 36.03 |

Table 12.8 Estimates of Land Area Affected by Strip Mining and Other Mining Activities

| State | Stripmines, Quarries | State | Stripmines, Quarries |
|---|---|---|---|
| | 1000 ha | | 1000 ha |
| Alabama | 49.2 | Nevada | 75.7 |
| Arizona | 62.2 | New Hampshire | 34.3 |
| Arkansas | 41.3 | New Jersey | 35.0 |
| California | 147.2 | New Mexico | 112.3 |
| Colorado | 72.4 | New York | 103.7 |
| Connecticut | 34.8 | North Carolina | 78.5 |
| Delaware | 0.9 | North Dakota | 42.7 |
| Florida | 294.4 | Ohio | 121.9 |
| Georgia | 59.5 | Oklahoma | 82.7 |
| Idaho | 53.6 | Oregon | 40.5 |
| Illinois | 101.2 | Pennsylvania | 443.3 |
| Indiana | 62.7 | Rhode Island | 5.3 |
| Iowa | 74.8 | South Carolina | 26.7 |
| Kansas | 103.3 | South Dakota | 24.2 |
| Kentucky | 253.6 | Tennessee | 83.5 |
| Louisiana | 43.5 | Texas | 300.8 |
| Maine | 43.3 | Utah | 25.2 |
| Maryland | 19.8 | Vermont | 17.7 |
| Massachusetts | 63.2 | Virginia | 46.9 |
| Michigan | 153.6 | Washington | 65.8 |
| Minnesota | 238.8 | West Virginia | 81.2 |
| Mississippi | 59.1 | Wisconsin | 95.0 |
| Missouri | 139.2 | Wyoming | 161.4 |
| Montana | 130.6 | Total | 4434.1 |
| Nebraska | 27.3 | | |

## 12.6 CONCLUSION

Mining causes severe or drastic soil disturbance. It alters vegetation and hydrology, and causes severe environmental problems. Because of the unsightly spoil banks and contamination and pollution of surface waters, several legislative policies have been developed to reclaim these drastically disturbed soils. Legislation for soil restoration was developed in West Virginia, Indiana, and Ohio in the 1940s and Kentucky in the 1950s (Plass, 2000). These laws established specific and stringent standards for reclamation. Surface mining was a political issue in the 1960s, and still is. Legislation was enacted in Illinois (1961) to achieve higher standards for compliance and stiffer penalties for noncompliance. Complete or partial leveling was made mandatory in most of the midwestern states. Additional laws were enacted in Virginia, Tennessee, and Alabama during the 1960s based on the first surface mining report (U.S. Department of Interior, 1967). The Federal Surface Mining Control and Reclamation Act of 1977 (PL95-87) requires that mine operators replace topsoil and spoil materials in an effort to duplicate the preexisting landscape and soils. These laws require regrading of the mine spoil to approximate pre-mining land contour, replacing the topsoil, and establishing the vegetative cover by the mining company (Barnhisel and Hower, 1997). Federal and state regulations specify design and performance standards to ensure that lands are restored to their approximate pre-mine use and/or level of productivity. With these regulations, the risks of adverse environment impact of soil disturbance by mining are minimized. The primary goal of reclamation is to reestablish a stable landscape with minimal risks of runoff, erosion, acid mine drainage, and pollution/contamination of natural waters.

## REFERENCES

Barnhisel, R.I. and J.M. Hower. Coal surface mine reclamation in the eastern U.S.: The revegetation of disturbed lands to hayland/pasture land or cropland. *Adv. Agron.* 61, 233–275, 1997.

Bastian, R.K., A. Montague, and T. Numbers. Surface mining. In *Land Reclamation and Biomass Production with Municipal Wastewater and Sludge.* W.E. Sopper, E.M. Speaker, and R.K. Bastian (Eds.), Penn State University Press, University Park, PA, 1982, pp. 13–54.

Chapman, E.W. Surface mining areas can be restored or left to ugliness. In *Using Our Natural Resources, 1983 Yearbook of Agriculture.* Washington, D.C., 1983, pp. 394–403.

Plass, W.T. History of surface mining and associated legislation. In *Reclamation of Drastically Disturbed Lands.* ASA/CSSA/SSSA Special Rep. 41, Madison, WI, 2000, pp. 1–19.

Sutton, P. and W.A. Dick. Reclamation of acid mined lands in humid areas. *Adv. Agron.* 41, 377–405, 1987.

U.S. Dept. of Interior. Surface mining and our environment. U.S. Government Printing Office, Washington, D.C., 1967.

# CHAPTER 13

# Wetland Degradation

A wetland "is an ecosystem that arises when inundation by water produces soils dominated by anaerobic processes and forces the biota, particularly rooted plants, to exhibit adaptations to tolerate flooding" (Keddy, 2000). Inherent in this definition are three important aspects of wetlands: inundation by water, reduction of the oxygen level in the soil, and predominance of biota that can tolerate an anaerobic environment. Therefore, the minimal essential characteristics of wetlands are: recurrent and sustained inundation that results in anaerobic conditions, hydric soils, and hydrophytic vegetation. Moreover, wetlands are transitional lands between terrestrial and aquatic ecosystems where the water table is at or near the surface throughout much of the year (Plate 13.1).

In the context of this volume, wetland degradation implies drainage of wetlands and their conversion to agricultural land use. To farmers, conversion of wetlands to agriculturally productive soils may not be soil degradation. In the environmental context, however, loss of wetlands is an important issue that needs to be addressed at regional, national, and international levels.

## 13.1 TYPES OF WETLANDS

Wetlands may be classified according to their hydrologic and vegetation characteristics or their geographical location. On the basis of their hydrologic and vegetation characteristics, there are six types of wetlands (Keddy, 2000; Mitsch and Gosselink, 2000). These are:

1. *Swamp:* A wetland ecosystem dominated by trees that are rooted in hydric soils but not in peat.
2. *Marsh:* A wetland ecosystem dominated by herbaceous plants rather than trees. Herbaceous plants are also rooted in hydric soils and not in peat. Some examples of herbaceous plants are cattail (*Typha* spp.) and reed (*Phragmites* spp.).
3. *Bog:* A wetland ecosystem dominated by *Sphagnum* moss, sedges, Ericaceous shrubs, or evergreen trees.

**Plate 13.1**   Most wetlands were drained and converted to agricultural land during the 18th and 20th centuries.

4. *Fen:* A wetland ecosystem dominated by sedges and grasses. The vegetation is rooted in a shallow peat. Fen is characterized by a considerable water movement through the peat.
5. *Wet meadow:* A wetland ecosystem dominated by herbaceous plants. The vegetation is frequently rooted in flooded soils. These wetlands are characterized by periodic flooding alternated by dry periods.
6. *Shallow water:* A wetland ecosystem dominated by aquatic plants growing in at least 25 cm of water.

On the basis of their geographical location, there are two types of wetlands (Mitsch and Gosselink, 2000). These are:

1. *Coastal wetlands:* These ecosystems comprise salt marsh, tidal freshwater marsh, mangroves, etc. Coastal wetlands in the lower 48 states and Alaska cover an area of 3.2 Mha. Alaskan wetlands cover an area of about 0.86 Mha (Hall et al. 1994) (Table 13.1).
2. *Inland wetlands:* These ecosystems comprise freshwater marsh, peatland, freshwater swamp, and riparian ecosystems. Inland wetlands in the lower 48 states and Alaska cover an area of 107 Mha (Table 13.1). Wetlands in the lower 48 states cover an area of 32 Mha to 38 Mha (Frayer et al., 1983; Mitsch and Gosselinks, 2000).

With the exception of about 2.3 Mha, most of the wetlands are inland wetlands (Mitsch and Gosselink, 2000). Wetlands may also be classified on the basis of permanence. According to these criteria, wetlands may be permanent, semipermanent, or temporary (Fredrickson, 1983). Wetland quality and type depend on soil,

**Table 13.1    Wetlands in the U.S. Including Alaska**

| Type of Wetland | Area (Mha) |
|---|---|
| Coastal wetlands | |
|    Tidal salt marshes | 1.9 |
|    Tidal freshwater marshes | 0.8 |
|    Mangrove wetlands | 0.5 |
| Inland wetlands | |
|    Freshwater marshes | 27 |
|    Peatlands | 55 |
|    Freshwater/riparian swamps | 25 |
| Total | 111 |

Note that area includes 68.8 Mha of wetlands in Alaska and 42.2 Mha in the contiguous 48 states.

*Source*: From Mitsch, W.J. and Gosselink, J.G., *Wetlands*, 3rd ed., John Wiley & Sons, New York, 2000.

water, duration and depth of flooding, and vegetation. Vegetation plays an important role in the quality of wetlands.

Extensive wetland areas in the U.S. are associated with river deltas. Important wetlands in the lower 48 states are the Mississippi Alluvial Plain and the associated ecosystems along the Gulf Coast, and the Sacramento River delta. In Alaska, wetlands are associated with the Copper, Kuskokwim, and Yukon River deltas (Fredrickson, 1983).

## 13.2  WETLAND SOILS

Wetland soils are predominantly anaerobic and deficient in oxygen leading to a hypoxia or anoxia environment. Wetlands provide the principal reducing system, leading to reduced condition. Consequently, the redox potential of wetlands may range from −300 to +350 mv. More details about the properties of wetland soils, especially in regard to biogeochemical cycles of elements, are given in Faulkner and Richardson (1989), Poonamperuma (1972; 1984), Richardson and Vepraskas (2001), and Mitsch and Gosselink (2000).

"Wetland soils," a general term for any soil found in a wetland, is often used interchangeably with "hydric soils." Hurt et al. (1998) defined hydric soils as those "formed under conditions of saturation, flooding or ponding long enough during the growing season to develop anaerobic conditions in the upper part." Most of the hydric soils are high in organic matter content and comprise Histosols and the Histel suborder of the Gelisols. The Gelisols are affected by permafrost and are mostly found in Alaska. Histosols cover about 2.9% of the U.S. (Collins and Kuehl, 2001).

The data in Table 13.2 show the distribution of Histosols by states. Total peatland area in the U.S. is 23.18 Mha, of which 15.3 Mha is Histosols. Alaska alone accounts for 56% of all peatlands in the U.S.

Table 13.2    Area of Organic Soils in the U.S.

| State | Area under Histosol (10³ ha) | State | Area under Histosol (10³ ha) |
|---|---|---|---|
| Alabama | 80.9 | Nebraska | 4.4 |
| Alaska | 4246.0 | Nevada | 7.4 |
| Arizona | 0 | New Hampshire | 89.9 |
| Arkansas | 0 | New Jersey | 73.2 |
| California | 61.7 | New Mexico | 0.1 |
| Colorado | 33.5 | New York | 313.1 |
| Connecticut | 43.4 | North Carolina | 633.9 |
| Delaware | 35.6 | North Dakota | 2.6 |
| Florida | 1594.3 | Ohio | 30.9 |
| Georgia | 187.9 | Oklahoma | 0 |
| Hawaii | 192.0 | Oregon | 32.9 |
| Idaho | 23.6 | Pennsylvania | 16.3 |
| Illinois | 35.6 | Puerto Rico | 2.8 |
| Indiana | 149.0 | Rhode Island | 11.9 |
| Iowa | 30.1 | South Carolina | 65.0 |
| Kansas | 0 | South Dakota | 0 |
| Kentucky | 0 | Tennessee | 0 |
| Louisiana | 953.7 | Texas | 5.2 |
| Maine | 396.5 | Utah | 2.8 |
| Maryland | 94.9 | Vermont | 27.0 |
| Massachusetts | 136.4 | Virginia | 54.9 |
| Michigan | 1651.1 | Washington | 79.0 |
| Minnesota | 2434.5 | West Virginia | 0 |
| Mississippi | 90.8 | Wisconsin | 1347.6 |
| Missouri | 5.1 | Wyoming | 3.0 |
| Montana | 26.0 | Total | 15306.5 |

*Source*: Modified from Bridgham, S.D. et al., in *Wetland Soils: Genesis, Hydrology, Landscape and Classification*, Lewis Publishers, Boca Raton, FL, 2001, 343–370.

Table 13.3    Estimates of Wetland Area in the Contiguous U.S.

| Period | Area (Mha) |
|---|---|
| Presettlement | 86.2–89.5 |
| 1900–1922 | 32–37 |
| 1940–1954 | 30.1–43.8 |
| 1970–1974 | 40.1–42.8 |
| 1980s | 41.8 |

The area of wetlands in Alaska (68.8 Mha) has remained the same.

*Source*: Modified from Mitsch, W.J. and Gosselink, J.G., *Wetlands*, 3rd ed., John Wiley & Sons, New York, 2000.

## 13.3  HISTORIC LOSS OF WETLANDS

Several reports indicate a rapid loss of wetlands prior to the 1970s (Dahl, 1990; Frayer et al., 1983; Tiner, 1984; Dahl and Johnson, 1991; Mitsch and Gosselink, 2000). The data in Table 13.3 show that in the mid-1950s, there were a maximum

Table 13.4   Some Examples of Wetland Losses

| Locale | Wetland Type | Area (Mha) Original | 1983 | % Loss | Principal Cause of Loss |
|---|---|---|---|---|---|
| Mississippi Delta | Lowland hardwood | 10.1 | 1.9 | 80.8 | Agriculture |
| Missouri | Lowland hardwood | 1.0 | 0.03 | 97.2 | Agriculture |
| Louisiana | Coastal marsh | 1.42 | 1.1 | 23.1 | Water development |
| North Central states | Prairie potholes | 51.4 | 28.3 | 44.9 | Agriculture |
| Texas | Playa Lakes | 0.10 | 0.02 | 79.8 | Agriculture |

Source: Modified from Fredrickson, L.H., in Using our Natural Resources, 1983 Yearbook of Agriculture. Washington, D.C., 234–243.

of 43.8 Mha of wetlands in the U.S. (Frayer et al., 1983). By the mid-1970s, however, the maximum area under wetlands in the U.S. was 42.8 Mha. The data in Table 13.3 do not show a clear trend with regard to temporal changes in the area under wetlands.

Data in Table 13.4 show some specific examples of the loss of wetland. The north central states and Mississippi had the greatest loss of this important natural resource. In addition to agriculture, wetland conversion was due to water resources development and urbanization.

The data in Table 13.5 show state-by-state estimates of wetlands. Since the wetland area in Alaska did not change, the area in the conterminous U.S. was 89.54 Mha in the 1780s and 42.30 Mha in the 1980s. It is apparent, therefore, that 53% of the wetlands in the conterminous U.S. were lost from the 1780s to the 1980s (Dahl, 1990). Drastic losses occurred in Florida, Texas, Ohio, Indiana, Illinois, and Minnesota (Table 13.5). Some studies estimated the loss rate of coastal wetlands at 8100 ha/yr between 1922 and 1954, and 19,000 ha/yr between 1954 and the 1970s (Gosselink and Baumann, 1980). Most of the wetland loss occurred due to logging and subsequent conversion to agriculture. Millions of hectares of our best agricultural lands were once wetland ecosystems.

## 13.4  CONSEQUENCES OF WETLAND LOSS

Wetlands are a vital natural resource. Wetlands provide flood protection, filter pollutants from runoff and seepage/drainage water from agricultural lands, improve water and air quality, and provide habitat for wildlife. The principal effects of wetland loss are on water quality, and decomposition of organic matter leading to emission of $CO_2$ into the atmosphere. The loss of wetlands in the upper Midwest has led to much more flooding downstream as the water is released much faster. Under natural conditions, wetlands are a net sink for C because they accumulate C in soil as peat or humus. When drained, however, former wetlands became a major source of C because of increase in the rate of mineralization. In addition to emission of $CO_2$ and other greenhouse gases, loss of wetlands also leads to pollution and contamination of natural waters. Wetlands are a natural filter for pollutants, especially the

Table 13.5   Loss of Wetlands in the U.S.

| State | Area (Mha) | | % loss |
|---|---|---|---|
|  | 1780 | 1980 |  |
| Alabama | 3.06 | 1.53 | 50.0 |
| Alaska | 68.80 | 68.80 | 0.0 |
| Arizona | 0.38 | 0.24 | 36.8 |
| Arkansas | 3.99 | 1.12 | 71.9 |
| California | 2.02 | 0.18 | 91.1 |
| Colorado | 0.81 | 0.41 | 49.4 |
| Connecticut | 0.27 | 0.07 | 74.1 |
| Delaware | 0.19 | 0.09 | 52.6 |
| Florida | 8.23 | 4.47 | 45.7 |
| Georgia | 2.77 | 2.14 | 29.4 |
| Hawaii | 0.02 | 0.02 | 100.0 |
| Idaho | 0.36 | 0.16 | 55.6 |
| Illinois | 3.32 | 0.51 | 2.81 |
| Indiana | 2.27 | 0.30 | 86.8 |
| Iowa | 1.62 | 0.17 | 89.5 |
| Kansas | 0.34 | 0.18 | 47.1 |
| Kentucky | 0.63 | 0.12 | 81.0 |
| Louisiana | 6.55 | 3.56 | 45.6 |
| Maine | 2.61 | 2.10 | 19.5 |
| Maryland | 0.67 | 0.18 | 73.1 |
| Massachusetts | 0.33 | 0.24 | 27.3 |
| Michigan | 4.53 | 2.26 | 50.1 |
| Minnesota | 6.10 | 3.52 | 42.3 |
| Mississippi | 4.00 | 1.65 | 58.8 |
| Missouri | 1.96 | 0.26 | 86.7 |
| Montana | 0.46 | 0.34 | 26.1 |
| Nebraska | 1.18 | 0.77 | 34.7 |
| Nevada | 0.20 | 0.10 | 50.0 |
| New Hampshire | 0.09 | 0.08 | 12.5 |
| New Jersey | 0.61 | 0.37 | 39.3 |
| New Mexico | 0.29 | 0.20 | 31.0 |
| New York | 1.04 | 0.42 | 59.6 |
| North Carolina | 4.49 | 2.30 | 48.8 |
| North Dakota | 1.99 | 1.01 | 49.2 |
| Ohio | 2.02 | 0.20 | 90.0 |
| Oklahoma | 1.15 | 0.38 | 67.0 |
| Oregon | 0.92 | 0.56 | 39.1 |
| Pennsylvania | 0.46 | 0.20 | 56.5 |
| Rhode Island | 0.04 | 0.03 | 25.0 |
| South Carolina | 2.60 | 1.89 | 27.3 |
| South Dakota | 1.11 | 0.72 | 35.1 |
| Tennessee | 0.78 | 0.32 | 59.0 |
| Texas | 6.48 | 3.08 | 59.0 |
| Utah | 0.33 | 0.23 | 52.5 |
| Vermont | 0.14 | 0.09 | 30.3 |
| Virginia | 0.75 | 0.44 | 35.7 |
| Washington | 0.55 | 0.38 | 41.3 |

(*continued*)

Table 13.5   Loss of Wetlands in the U.S. (*Continued*)

| State | Area (Mha) | | % loss |
|---|---|---|---|
| | 1780 | 1980 | |
| West Virginia | 0.05 | 0.04 | 30.9 |
| Wisconsin | 3.97 | 2.16 | 20.0 |
| Wyoming | 0.81 | 0.51 | 45.6 |
| Total | 158.34 | 111.10 | 37.0 |
| Total in conterminous U.S. | 89.54 | 42.30 | 29.8 |

*Source*: Modified from Dahl, T.E., *Wetlands Losses in the United States, 1970s to 1980s*, 1990; Mitsch, W.J. and Gosselink, J.G., *Wetlands*, 3rd ed., John Wiley & Sons, New York, 2000.

agricultural chemicals. Loss of wetlands increases the risk of flooding without contaminants being filtered and retained. In the absence of wetlands, contaminants from agricultural lands are directly transported into aquatic ecosystems. The severe problem of hypoxia in the Gulf of Mexico is linked to the loss of wetlands in the Mississippi Basin.

The NRI data showed that an average of 21,862.3 ha of wetland area was lost between 1992 and 1997 from conversion to cropland or pastureland. During the same period, an average of 12,145.7 ha was gained by the conversion of crop and pastureland back to wetlands through the wetland reserve program (WRP) (NRCS, 2000). The net loss of wetlands on these lands decreased from a level of 10,931.2 ha lost each year (1982–1992 period) to 9716.6 ha from 1992 to 1997.

## REFERENCES

Bridgham, S.D., C.-L. Ping, J.L. Richardson, and K. Updagraff. Soils of northern peatlands: Histosols and Gelisols. In *Wetland Soils: Genesis, Hydrology, Landscape and Classification*. J.L. Richardson and M.J. Vepraskas (Eds.), Lewis Publishers, Boca Raton, FL, 2001, pp. 343–370.

Collins, M. and J.R. Kuehl. Organic matter accumulation and organic soils. In *Wetland Soils: Genesis, Hydrology, Landscape and Classification*. J.L. Richardson and M.J. Vepraskas (Eds.), Lewis Publishers, Boca Raton, FL, 2001, pp. 137–161.

Dahl, T.E. Wetlands losses in the United States, 1780s to 1980s. U.S. Fish and Wildlife Service, Washington, D.C.,1990.

Dahl, T.E. and C.E. Johnson. Wetlands status and trends in the conterminous United States, mid-1970s to mid-1980s. U.S. Dept. of Interior, Fish and Wildlife Service, Washington, D.C., 1991, 28 pp.

Faulkner, S.P. and C.J. Richardson. Physical and chemical characteristics of freshwater wetland soils. In *Constructed Wetlands for Wastewater Treatment: Municipal, Industrial and Agricultural*. D.A. Hammer (Ed.), Lewis Publishers, Chelsea, MI, 1989, pp. 41–72.

Frayer, W.E., T.J. Monahan, D.C. Bowden, and F.A. Graybill. Status and trends of wetlands and deepwater habitat in the conterminous United States, 1950s to 1970s. Department of Forest and Wood Sciences, Colorado State University, Fort Collins, CO, 1983, 32 pp.

Fredrickson, L.H. Wetlands: a vanishing resource. In *Using our Natural Resources, 1983 Yearbook of Agriculture*. USDA Government Printers, Washington, D.C., 1983, pp. 234–243.

Gosselink, J.G. and R.H. Baumann. Wetland inventories: Wetland loss along the United States coast. *Z. Geomorphol. N.F. Suppl. Bd.* 34, 173–187, 1980.

Hall, J.V., W.E. Frayer, and B.O. Wilen. Status of Alaska Wetlands. U.S. Fish and Wildlife Service, Alaska Region, Anchorage, 1994, 32 pp.

Hurt, G.W., P.M. Whited, and R.F. Pringle (Eds.). Field indicators of hydric soils in the U.S. USDA-NRCS, Fort Worth, TX, 1998.

Keddy, P.A. *Wetland Ecology: Principles and Conservation*. Cambridge University Press, Cambridge, U.K., 2000, 614 pp.

Mitsch, W.J. and J.G. Gosselink. *Wetlands*, 3rd ed. John Wiley & Sons, New York, 2000.

NRCS. National Resource Inventory: Background and Highlights. USDA-NRCS, Washington, D.C., 2000.

Poonamperuma, F.N. The chemistry of submerged soils. *Adv. Agron.* 24, 29–96, 1972.

Poonamperuma, F.N. Effects of flooding on soils. In *Flooding and Plant Growth*. T. Kozlowski (Ed.), Academic Press, Orlando, FL, 1984, pp. 9–45.

Richardson, J.L. and M.J. Vepraskas. *Wetland Soils: Genesis, Hydrology, Landscape and Classification*. Lewis Publishers, Boca Raton, FL, 2001, 409 pp.

Tiner, R.W. Wetlands of the United States: Current status and recent trends. National Wetlands Inventory, U.S. Fish and Wildlife Service, Washington, D.C., 1984, 58 pp.

# SECTION IV

# Policy and Conservation Programs

# CHAPTER 14

# Policy Options

The fundamental cause of soil degradation is public attitude toward and perception of the overall environment and how natural resources need to be managed in a sustainable way. The lack of a basic understanding of soils is further confounded and exacerbated by the lack of knowledge with regard to the impact of anthropogenic perturbations on soil properties and processes. Taking soil for granted (e.g., it is there for human use, it can withstand abuse, its capacity to produce is limitless, degradation is temporary, and technology can fix or solve any problem) can exacerbate the problem. Also, the failure to see what the impacts of farming and other operations have on soils is not understood, yields have continued to increase even as the natural soil resource was degraded because of the increasing use of inputs and adoption of improved varieties. It was because of this masking effect of inputs on crop yields that the adverse impacts on soil degradation and decline in air and water qualities were not comprehended. Therefore, controlling soil degradation and restoring degraded soils require behavioral change to alter the attitudes and perceptions. The human dimensions of soil degradation are complex, but understanding these dimensions is important to establishing appropriate policies that minimize risks and restore degraded soils. It was not until 1972 that a policy was put in place that required restoration of mined soils.

There have been significant advances in soil sciences and in agricultural technologies which, when adopted, can restore degraded soils and facilitate sustainable management of natural resources. Soil management technology exists to prevent, arrest and restore most of the degraded soils (Plate 14.1 and Plate 14.2). For example, much of the observed decrease in soil erosion during the 1990s was caused by conversion of highly erodible land (HEL) to the Conservation Reserve Program (CRP). To do so, however, would require identification and implementation of appropriate policies to ensure that soils are managed in such a way as to maintain their productivity and environment moderation capacity indefinitely. The beneficial assets or properties of the soil for numerous functions of value to humans must be preserved, restored, and enhanced. These assets include topsoil depth, SOC pool, soil structure, available water and nutrient retention and supply capacities, ability

**Plate 14.1**   Conservation tillage decreases risks of soil erosion and sequesters soil carbon.

**Plate 14.2**   Contour farming and buffer strips are conservation-effective measures.

to resist degradation or soil's resilience against erosion and salinization. Among numerous degradation processes, accelerated erosion remains the most widespread and highly deleterious. In addition to decline in crop yield, impairment and pollution of natural waters by sedimentation and eutrophication is the most severe off-site impact (NRC, 1993; NRCS, 1993; 1996; Foster and Dabney, 1995; Crosson and Anderson, 2000).

## 14.1 HISTORY OF SOIL CONSERVATION PROGRAMS IN THE U.S.

Soil conservation programs in the U.S. have existed ever since the Dust Bowl era of the 1930s (Table 14.1) (Braeman, 1986; Worster, 1979). Several policies have been put in place to induce voluntary adoption of conservation practices. State laws were put in place to implement the conservation district (county) program designed

Table 14.1   Soil and Water Conservation Policies within the U.S.

| Year | Provision/Act | Objective |
|------|---------------|-----------|
| 1933 | Soil Erosion Service within the Dept. of Interior | To conduct research on the causes and consequences of erosion |
| 1935 | Soil Conservation Service within USDA | To provide technical assistance and conduct research |
| 1936 | Soil Conservation and Domestic Allotment Act | To legitimize government payments and provide economic incentives to farmers |
| 1944–1945 | Agricultural Conservation Program (ACP) | To divert cropland from production and retire land |
| 1956 | Soil Bank Program | To address soil erosion and water pollution problems |
| 1985 | Food Security Act | To require landowners to comply with contractual agreements or suffer loss of access to federal farm programs |
|      | • Conservation Reserve Program (CRP) | To retire HEL cropland from production by government purchase of cropping rights for the duration of lease agreements |
|      | • Conservation Compliance (T-2000) | To mandate owners of HEL to develop and implement conservation plan by 1-1-95 or pay penalties |
| 1996 | Federal Agricultural Improvement and Reform (FAIR) Act | To implement soil and water conservation initiatives |
|      | • Environment Quality Incentive Program (EQUIP) | To implement CRP and WRP |
|      | • Environment Conservation Acreage Reserve Program (ECARP) | To focus conservation programs on watersheds, multi-state regions or regions of special environment sensitivity, and maximize environmental benefits |
| 2002 | • Comprehensive Conservation Enhancement Program | • Implement soil and water conservation initiatives |
|      | • Environment Quality Incentive Program | • Implement CRP, WRP, and grassland reserve initiatives |
|      | • Farmland Protection Program | • Resource Conservation and Development Program |

*Source*: Taken from Rasmussen, W.D. and Baker, G.L., USDA, Washington, D.C., 1979; Harmon, K.W., in *Impacts of the Conservation Reserve Program in the Great Plains*, J.E. Mitchell (Ed.), USDA-FS, Denver, CO, 1987, pp. 105–108; Rasmussen, W.D., in *Soil Conservation Policies, Institutions and Incentives*, H.G. Halcrow, E.O. Heady, and M.L. Cotner (Eds.), Soil Water Cons. Soc., Ankeny, IA, 1982, pp. 3–18; Helms, D., in *Implementing the Conservation Title of the Food Security Act of 1985*, T.L. Napier (Ed.), Soil Water Cons. Soc., Ankeny, IA, 1990, pp. 11–25; Napier, T.L. and Napier, S.M., in *Soil and Water Conservation Policies and Programs: Successes and Failures*, T.L. Napier, S.M. Napier, and J. Turdon (Eds.), CRC Press, Boca Raton, FL, 2000, pp. 83–93.

to provide education and technical assistance to farmers (Libby, 1982). Agricultural Stabilization and Conservation (ASC) committees were established at the county level to handle cost sharing. Funds were made available through the Agricultural Conservation Program (ACP) to implement practices (Holmes, 1987). Some of these policies, implemented by the USDA, include on-farm technical assistance, extension education, cost-sharing assistance for installing preferred practices, rental and ease-ment payments to take land out of production and place it in conservation use (Uri and Lewis, 1998).

### 14.1.1  Food Security Act of 1985

A major landmark of conservation policy in the U.S. was the Food Security Act of 1985 (Heimlich, 1991). This act provides farmers the economic incentive to adopt conservation tillage and other practices, and links agricultural program payments to the adoption of an acceptable conservation system on HEL. To receive agricultural payments on HEL-designated cropland, a farmer has to implement an acceptable conservation plan (Crosswhite and Sandretto, 1991).

### 14.1.2  Conservation Reserve Program (CRP)

A forerunner of the CRP was the Soil Bank Program of the 1950s. The CRP, as a part of the Food Security Act of 1985, enables the USDA to enter into a 10- to 15-year agreement with land owners (operators) to remove HEL and other environ-mentally sensitive cropland from production (Osborn, 1996). The CRP has been extremely effective in reducing the sediment load (Magleby et al., 1995), improving soil quality, and sequestering C in soil and the ecosystem (Gebhart et al., 1994; Barker et al., 1996; Follett et al., 2001).

### 14.1.3  Federal Agricultural Improvement and Reform (FAIR) Act of 1996

The FAIR Act modifies the conservation compliance provisions of the Food Security Act of 1985. The FAIR Act provides farmers with greater flexibility in developing and implementing conservation plans. Violators of implementing conser-vation plans on HEL risk loss of eligibility for many payments. An important provision of the FAIR Act is self-certifying compliance (Nelson and Schertz, 1996; Uri, 2001).

### 14.1.4  The Environmental Quality Incentive Program (EQUIP)

The EQUIP is a special provision of the FAIR Act of 1996. It makes provisions for farmers to adopt production practices that reduce environmental and resource problems. It encourages implementation of CRP and WRP. The FAIR Act contains elements of cross-compliance that will penalize farmers and land owners for non-compliance.

## 14.1.5  Environmental Conservation Acreage Reserve Program (ECARP)

The ECARP is another provision of the FAIR Act of 1996. Activities under ECARP focus on watersheds, multi-state regions, and other regions of environmental sensitivity. The goal of ECARP is to maximize environmental benefits and to facilitate adoption of conservation effective measures. Most of the policies outlined in Table 14.1 are of two types. The first is to set aside completely or retire land from production. In this case, the land is put into a conservation cover. The second is to reduce erosion levels from managed ecosystems. The second requires special equipment, appropriate cropping systems, judicious management, or a combination of these options. Implementation of these policies has relied on several tools such as technical assistance, conservation cost share, buy-outs, and set-asides (Hoag et al., 2000). Since implementation of these policies in the 1930s, the question has been whether they have been successful.

## 14.2  IMPACT OF CONSERVATION POLICIES IN THE U.S.

The data presented in this book indicate that these policies have been effective in reducing the extent, magnitude, and rates of soil degradation by different processes. Specifically, there have been reductions in soil degradation between 1982 and 1997 as follows:

- Total erosion on cropland decreased by 42% between 1982 and 1997 from 3.1 billion Mg/yr in 1982 to 1.8 billion Mg/yr in 1997.
- Total erosion on HEL cropland fell by 46.5% between 1982 and 1997 from 1.7 billion Mg in 1982 to 0.91 billion Mg in 1997.
- Sheet and rill erosion on all cropland decreased by 38.4% between 1982 and 1997 from 1.7 billion Mg in 1982 to 1.0 billion Mg in 1997.
- Combined sheet and rill erosion on HEL cropland decreased by 46.7% between 1982 and 1997 from 0.91 billion Mg in 1982 to 0.45 billion Mg in 1997.
- Wind erosion on all cropland decreased by 46.3% from 1.4 billion Mg/yr in 1982 to 1.0 billion Mg/yr in 1997.
- Wind erosion on HEL cropland also decreased by 46.3% from 0.8 billion Mg in 1982 to 0.45 billion Mg in 1997.
- Soil erosion on all cropland decreased from a rate of 17.9 Mg/ha/yr in 1982 to 11.7 Mg/ha/yr in 1997.
- Sheet and rill erosion on all cropland decreased from 9.9 Mg/ha/yr in 1982 to 6.7 Mg/ha/yr in 1997. Wind erosion on cropland decreased from 8.1 Mg/ha/yr in 1982 to 4.9 Mg/ha/yr in 1997.
- Erosion on HEL cropland decreased from 33.9 Mg/ha/yr in 1982 to 20.9 Mg/ha/yr in 1997. Sheet and rill erosion on HEL cropland decreased from 18.4 Mg/ha/yr in 1982 to 11.4 Mg/ha/yr in 1997. Wind erosion on HEL cropland decreased from 15.5 Mg/ha/yr in 1982 to 9.4 Mg/ha/yr in 1997.
- Erosion on NHEL cropland decreased from 11.2 Mg/ha/yr in 1982 to 7.8 Mg/ha/yr in 1997. Sheet and rill erosion on NHEL cropland decreased from 6.3 Mg/ha/yr in 1982 to 4.7 Mg/ha/yr in 1997. Wind erosion on NHEL cropland decreased from 4.9 Mg/ha/yr in 1982 to 3.1 Mg/ha/yr in 1997.

These data show significant reduction in total erosion and the rate of erosion on all land types, especially cropland in the U.S. between 1982 and 1997. This reduction is indicative of the success of soil conservation policies. The reduction was most pronounced on HEL cropland, which was the target of policies implemented since the 1950s. Yet, the problem is still a serious issue. Addressing the problem that remains requires further policy considerations especially with regard to the off-site impacts of erosion.

## 14.3 FUTURE POLICY OPTIONS

The policies chosen since the 1930s have used four tools to attain their goals: technical assistance, conservation cost share, buy-outs, and set-asides (Jeffords, 1982; Laycock, 1991; Riechelderfer, 1991; NRC, 1993). These policy tools amount to financial incentives. Regulations have rarely been applied and financial disincentives (erosion tax) have never been used (Riechelderfer, 1991). These policies alone may not be adequate to decrease further losses, because of the countereffects of incentives provided in other programs (Hoag et al., 2000). Numerous commodity programs reward farmers for high yields, which may exacerbate erosion on some soils. Besides, none of these policies have considered the societal benefits or off-site benefits of soil conservation, especially with regard to water quality and soil carbon sequestration.

### 14.3.1 Emphasis on Water Quality and Soil C Sequestration

The CRP program has been politically popular and a successful undertaking. Therefore, CRP will be continued with the emphasis being on water quality and soil C sequestration rather than on soil saving. In addition to incentives, however, regulations and financial disincentives may also be used whenever it is more profitable to farmers to employ production systems to threaten soil and water resources in environmentally sensitive ecoregions. Therefore, it is likely that more emphasis will be placed on regulatory approaches to motivate land managers in adopting specific measures with long-term societal benefits (Napier and Napier, 2000b). However, it is expensive to use regulatory measures, especially when monitoring individual farmsteads is required in implementing such policies.

Soil C sequestration may be an important target of any future policies. To be an effective policy tool, methodology for monitoring soil C sequestration must be in place. Such a methodology must be simple, transparent, and, above all, cost effective. It is also important to commodify soil carbon, so that it can be traded (bought and sold) like any other farm commodity. The beneficiaries of soil C sequestration may also include the overseas consumers. International consumers may pay the price of C sequestration through a "global market" for C trading.

Some of the policy tools that may be used to control erosion, improve water quality, and sequester soil C may include the following: purchasing cropping rights to HEL, permanent retirement of HEL, and cultivation of some soils with conservation tillage only. It may be necessary to accept and implement the polluter pays principle.

### 14.3.2  Private Sector Involvement

The private sector may play an important role in future policies dealing with environmental issues such as soil C sequestration to mitigate the greenhouse effect. Throwing money as subsidies may not necessarily solve the problem on a long-term basis. Future public policies may encourage strong participation by private sectors and may discourage government involvement in implementation of conservation programs. Future policies may also provide greater responsibility on landowners, who may be required to demonstrate that the production system in use is environmentally friendly and benign. This concept may involve user pays principle. This is in contrast to policies implemented between the 1930s and 1990s (Table 14.1) that required the state to prove that specific farmers were not in compliance with environmental standards (Napier and Napier, 2000b).

Rather than heavy emphasis on land stewardship (e.g., environmental ethics, values, intergeneration equity), future policies will emphasize business transactions and profit margins. Landowners will adopt alternative production systems if these are more profitable than those presently being used. Implementation of such a profit-based approach to conservation may involve the private sector. Encouraging emergence of private business to undertake implementation of an environmentally benign program may be a better approach than existing extension functions.

### 14.3.3  Recycling Biosolids

Greater emphasis will also be needed on capturing and recycling wastes. Some of the so-called wastes can be an important input to achieving soil conservation and enhancing the environment. If wastes are not captured and recycled, they can lead to pollution and contamination of natural waters and emission of GHGs into the atmosphere. Biosolids (e.g., crop residue, farm yard manure, forestry by-products) are important resources that must be used judiciously to conserve soil and water, sequester C, and recycle mineral elements.

### 14.3.4  Holistic Planning

Soil conservation will need to be linked to other issues (e.g., acid rain, preserving biodiversity, mitigating greenhouse effect, improving water quality). A piecemeal approach can be successful only on the short-term horizon. Any long-term solution must be based on a holistic approach that considers on-site and off-site impacts, alternative land uses, intergenerational issues, etc. Any segmental planning based on arbitrary boundaries (Johnson and Lewis, 1995) can only be partially successful.

### 14.3.5  Protecting HEL and Restoring Degraded Soils

It is important to protect HEL and restore soils degraded by prior misuse and mismanagement. It is likely that other ecosystems cannot survive unless HELs are protected because these are critical to the off-site adverse impacts. In addition to demarcating HEL, it may be appropriate to select and retain buffer zones between HEL and prime agricultural land.

### 14.3.6 Long-Term Perspective

It is important to keep the long-term sustainability in proper perspective, and balance long-term needs with short-term gains.

## 14.4 CONCLUSIONS

Policies implemented between the 1930s and 1990s have been successful in drastically reducing soil erosion from cropland and HEL. However, further reduction in erosion and soil degradation by other processes will require a newer approach. Controlling soil degradation requires long-term solutions that include a holistic approach, involvement of private sector, transferring responsibility to landowners, and adopting the user pays principle. Some of the subsidy programs are viewed as honey pots, providing short-term incentives. Long-term solutions must be based on alternative land uses that are profitable and involve a strong private sector rather than government agencies.

## REFERENCES

Barker, J., G. Baumgardner, D. Turner, and J. Lee. Carbon dynamics of the Conservation and Wetland Reserve Program. *J. Soil Water Cons.* 51, 340–346, 1996.
Braeman, J. The Dust Bowl: An introduction. *Great Plains Q.* 6, 67–68, 1986.
Crosson, P. and J.R. Anderson. Land degradation and food security: Economic impacts of watershed degradation. In *Integrated Watershed Management in the Global Ecosystem.* R. Lal (Ed.), CRC Press, Boca Raton, FL, 2000, pp. 291–303.
Crosswhite, W. and C. Sandretto. Trends in resource protection policies in agriculture. Agricultural Resources: Cropland, Water and Conservation Situation and Outlook Report. USDA-ERS, Washington, D.C., 1991.
Follett, R.F., E. Samson-Liebig, J.M. Kimble, E.G. Pruessner, and S. Waltman. Carbon sequestration under CRP in the historic grassland soils of the USA. In *Soil Carbon Sequestration and the Greenhouse Effect.* R. Lal (Ed.), SSSA Spec. Publ. #57, Madison, WI, 2001, pp. 27–39.
Foster, G. and S. Dabney. Agricultural tillage systems: water erosion and sedimentation. *Farming for a Better Environment.* Soil Water Cons. Soc., Ankeny, IA, 1995.
Gebhart, D., H. Johnson, H. Mayeux, and H. Polley. The CRP increases soil organic carbon. *J. Soil Water Cons.* 49, 488–492, 1994.
Harmon, K.W. History and economics of Farm Bill legislation and its impact on wildlife management and policies. In *Impacts of the Conservation Reserve Program in the Great Plains.* J.E. Mitchell (Ed.), USDA-FS, Denver, CO, 1987, pp. 105–108.
Heimlich, R. Soil erosion and conservation policies in the United States. In *Farming and the Countryside: An Economic Analysis of External Costs and Benefits.* N. Hanley (Ed.), CAB International, Wallingford, U.K., 1991.

Helms, D. New authorities and new roles: SCS and the 1985 Farm Bill. In *Implementing the Conservation Title of the Food Security Act of 1985.* T.L. Napier (Ed.), Soil Water Cons. Soc., Ankeny, IA, 1990, pp. 11–25.

Hoag, D.L., J.S. Hughes-Popp, and P.C. Huszar. Is U.S. soil conservation policy a sustainable development? In *Soil and Water Conservation Policies and Programs: Successes and Failures.* T.L. Napier, S.M. Napier, and J. Turdon (Eds.), CRC Press, Boca Raton, FL, 2000, pp. 127–142.

Holmes, B. Legal Authorities for Federal (USDA), State and Local Soil and Water Conservation Activities: Background for the Second RCA Appraisal. USDA, Washington, D.C., 1987.

Jeffords, J.M. Soil conservation policy for the future. *J. Soil Water Cons.* 37, 10–13, 1982.

Johnson, D.L. and L.A. Lewis. *Land Degradation: Creation and Destruction.* Blackwell, Oxford, U.K., 1995, 335 pp.

Laycock, W. The Conservation Reserve Program: How did we get where we are and where do we go from here? In *The Conservation Reserve—Yesterday, Today and Tomorrow.* L. Joyce, J. Mitchell, and M. Skold (Eds.), General Technical Report RM-203, USDA-FS, Rocky Mountain Forest and Range Experiment Station, Fort Collins, CO, 1991.

Libby, L. Interaction of RCA with state and local conservation programs. In *Soil Conservation, Policies, Institutions and Incentives.* H. Halcow, E. Heady, and M. Cotner (Eds.), Soil Cons. Soc. Am., Ankeny, IA, 1982.

Magleby, R., C. Sandretto, W. Crosswhite, and T. Osborn. Soil Erosion and Conservation in the U.S. USDA-ERS, AIB-718, Washington, D.C., 1995.

Napier, T.L. and S.M. Napier. Soil and water conservation policies in the U.S. In *Soil and Water Conservation Policies and Programs: Successes and Failures.* T.L. Napier, S.M. Napier, and J. Turdon (Eds.), CRC Press, Boca Raton, FL, 2000a, pp. 83–93.

Napier, T.L. and S.M. Napier. Future soil and water conservation and policies and programs within the U.S. In *Soil and Water Conservation Policies and Programs: Successes and Failures.* T.L. Napier, S.M. Napier, and J. Turdon (Eds.), CRC Press, Boca Raton, FL, 2000b, pp. 95–107.

National Research Council Soil and Water Quality. National Academy Press, Washington, D.C., 1993.

National Resource Conservation Service. USDA-NRCS, Washington, D.C., 1993.

National Resource Conservation Service. A Geography of Hope. USDA-NRCS, Washington, D.C., 1996.

Nelson, G. and L. Schertz. Provisions of the Federal Agriculture Improvement and Reform Act of 1996. USDA-ERS, Washington, D.C., 1996.

Osborn, T. Conservation: Provisions of the Federal Agricultural Improvement Act of 1996. USDA-ERS, Washington, D.C., 1996.

Rasmussen, W.D. History of soil conservation: institutions and incentives. In *Soil Conservation Policies, Institutions and Incentives.* H.G. Halcrow, E.O. Heady, and M.L. Cotner (Eds.), Soil Water Cons. Soc., Ankeny, IA, 1982, pp. 3–18.

Rasmussen, W.D. and G.L. Baker. Price support and adjustment program from 1933 through 1978: A short history. USDA, Washington, D.C., 1979.

Riechelderfer, K. The expanding role of environmental interest in agricultural policy. *World Agric.* 40, 18–21, 1991.

Uri, N. The environmental implications of soil erosion in the United States. *Environ. Monitoring Assess.* 66, 293–312, 2001.

Uri, N.D. and J.A. Lewis. The dynamics of soil erosion in U.S. agriculture. *Sci. Total Environ.* 218, 45–58, 1998.

Worster, D. *Dust Bowl.* Oxford University Press, New York, 1979.

# Appendices

| | Conventional Units | Metric Units | Factor for Converting from Conventional to Metric Units |
|---|---|---|---|
| Length | Mile (mi) | Kilometer (km) | 1.609 |
| | Yard (yd) | Meter (m) | 0.914 |
| | Foot (ft, ') | Meter (m) | 0.304 |
| | Inch (in, ") | Millimeter (mm) | 25.4 |
| Area | Acre | Hectare (ha, $10^4$ m$^2$) | 0.405 |
| | Square mile | Square kilometer (km$^2$) | 2.590 |
| | Square foot | Square meter (m$^2$) | $9.29 \times 10^{-2}$ |
| Volume | Cubic foot (ft$^3$) | Cubic meter (m$^3$) | $2.83 \times 10^{-2}$ |
| | Bushel (bu) | Liter (L) | 35.24 |
| | Quart (qt) | Liter (L) | 0.946 |
| | Pint (pt) | Cubic meter (m$^3$) | 0.0004732 |
| Mass | Pound (lb) | Gram (g) | 454 |
| | Ounce (oz) | Kilogram (kg) | 28.4 |
| | Quintal (q) | Kilogram (kg) | 100 |
| | Slug | Kilogram (kg) | 14.59 |
| | Short ton (200 lb, ton) | Kilogram (kg) | 907 |
| | Short ton (ton) | Megagram (Mg) | 0.907 |
| Yield | Pounds per acre (lb/acre) | Kilogram per hectare (kg/ha) | 1.12 |
| | Bushels per acre of soybean (60 lb/acre) | Megagram per hectare (Mg/ha) | 0.0673 |
| | Bushels per acre of corn (56 lb/acre) | Megagram per hectare (Mg/ha) | 0.0628 |
| | Bushels per acre of wheat (60 lb/acre) | Megagram per hectare (Mg/ha) | 0.0673 |
| | Bushels per acre of rice (45 lb/acre) | Megagram per hectare (Mg/ha) | 0.0505 |
| | Gallon per acre | Liter per hectare (L/ha) | 9.35 |

(*continued*)

**Appendix 1   Conversion of Units (*Continued*)**

|  | Conventional Units | Metric Units | Factor for Converting from Conventional to Metric Units |
|---|---|---|---|
| Temperature | Fahrenheit (ºF) | Celsius (ºC) | 0.556 (ºF–32) |
|  | Kelvin (K) | Celsium (ºC) | K–273 |
| Flow | Acre-inch (acre-in) | Cubic meter (m³) | 102.8 |
|  | Cubic foot per second (ft³/sec) | Cubic meter per hour (m³/hr) | 101.9 |
|  | Gallon per minute (gal/min) | Cubic meter per hour (m³/hr) | 0.227 |
|  | Acre-foot (acre-ft) | Hectare meter (ha m) | 0.123 |
| Nutrients in fertilizer | Phosphate (P₂O₅) | Elemental P | 0.437 |
|  | Potash (K₂O) | Elemental K | 0.830 |
|  | Lime (CaO) | Elemental Ca | 0.715 |
|  | Dolomite (MgO) | Elemental Mg | 0.602 |
| Pressure and force | Pounds per square foot (psf) | Pascal (Pa) | 4.88 |
|  | Pounds per square inch (psi) | Pascal (Pa) | 6894.766 |
|  | Pound (lb) | Newton (N) | 4.448 |
|  | Pound-foot (lb-ft) | Newton-meter (Nm) | 1.356 |
|  | One pound per cubic foot (lb/ft³) | Newton per cubic meter (N/m³) | 157.09 |
|  | Slug per cubic foot | Kilogram per cubic meter (kg/m³) | 515.4 |

**Appendix 2   Extent of Water Erosion for Cultivated Cropland by States
(Computed by Universal Soil Loss Equation)**

| State | <0.01 | ≥0.01 and <400.0 | ≥400.0 | Total |
|---|---|---|---|---|
| | | 1000 ha | | |
| Alabama | 10096.9 | 3294.7 | 0.0 | 13391.6 |
| Arizona | 15995.1 | 13531.1 | 0.0 | 29526.2 |
| Arkansas | 8050.5 | 5725.0 | 0.0 | 13775.5 |
| California | 29549.2 | 11555.9 | 0.0 | 41105.1 |
| Colorado | 12620.7 | 14339.0 | 0.0 | 26959.7 |
| Connecticut | 1129.2 | 170.5 | 0.0 | 1299.7 |
| Delaware | 306.4 | 223.1 | 0.0 | 529.5 |
| Florida | 11239.6 | 3954.3 | 0.0 | 15194.0 |
| Georgia | 11474.9 | 3782.8 | 0.0 | 15257.7 |
| Idaho | 15776.8 | 5866.5 | 0.0 | 21643.3 |
| Illinois | 3198.5 | 11393.8 | 0.1 | 14593.4 |
| Indiana | 2815.8 | 6556.2 | 0.0 | 9372.1 |
| Iowa | 1846.5 | 12728.9 | 0.0 | 14575.4 |
| Kansas | 1848.1 | 19461.8 | 0.0 | 21310.0 |
| Kentucky | 5625.7 | 4823.1 | 17.3 | 10466.1 |
| Louisiana | 8808.6 | 3559.1 | 0.0 | 12367.7 |
| Maine | 8234.2 | 381.5 | 0.0 | 8615.7 |
| Maryland | 1763.6 | 945.2 | 0.4 | 2709.2 |
| Massachusetts | 1940.9 | 204.7 | 0.0 | 2145.6 |
| Michigan | 10319.0 | 4839.6 | 0.0 | 15158.6 |
| Minnesota | 11248.2 | 10612.0 | 0.0 | 21860.2 |
| Mississippi | 7973.2 | 4378.4 | 0.0 | 12351.6 |
| Missouri | 6867.3 | 11182.2 | 2.2 | 18051.7 |
| Montana | 15331.4 | 22753.7 | 0.0 | 38085.1 |
| Nebraska | 1444.1 | 18591.0 | 0.0 | 20035.1 |
| Nevada | 25151.4 | 3484.0 | 0.0 | 28635.4 |
| New Hampshire | 2265.8 | 137.4 | 0.0 | 2403.2 |
| New Jersey | 1642.4 | 374.2 | 0.3 | 2016.9 |
| New Mexico | 14317.1 | 17175.6 | 0.0 | 31492.7 |
| New York | 8970.2 | 3747.8 | 1.0 | 12719.0 |
| North Carolina | 10234.2 | 3407.2 | 0.0 | 13641.4 |
| North Dakota | 1286.0 | 15812.0 | 0.0 | 18312.0 |
| Ohio | 4517.6 | 6185.4 | 1.3 | 10704.3 |
| Oklahoma | 4610.3 | 13508.3 | 0.0 | 18118.7 |
| Oregon | 18889.8 | 6252.3 | 0.0 | 25142.0 |
| Pennsylvania | 8105.5 | 3626.0 | 3.3 | 11734.9 |
| Rhode Island | 288.0 | 26.0 | 0.0 | 314.0 |
| South Carolina | 6158.5 | 1899.0 | 0.7 | 8058.2 |
| South Dakota | 2518.8 | 17454.2 | 0.0 | 19973.1 |
| Tennessee | 6507.1 | 4405.7 | 2.5 | 10915.3 |
| Texas | 10708.1 | 58395.3 | 0.0 | 69103.4 |
| Utah | 16536.1 | 5453.0 | 0.0 | 21989.0 |
| Vermont | 2080.8 | 408.9 | 0.3 | 2490.0 |
| Virginia | 7839.7 | 2718.9 | 0.0 | 10558.6 |
| Washington | 11845.8 | 5802.1 | 0.0 | 17647.9 |
| West Virginia | 5105.2 | 1160.4 | 10.3 | 6276.0 |
| Wisconsin | 8640.1 | 5903.6 | 0.0 | 14543.7 |
| Wyoming | 13672.0 | 11660.8 | 0.0 | 25332.8 |
| Total | 398609.2 | 383852.3 | 40.7 | 782502.1 |

**Appendix 3   Severity of Wind Erosion on Cultivated Cropland for Different States in the Conterminous U.S. (Computed by Wind Erosion Equation)**

| State | <0.01 | ≥0.01 and <400.0 | ≥400.0 | Total |
|---|---|---|---|---|
| | | 1000 ha | | |
| Alabama | 13391.6 | 0.0 | 0.0 | 13391.6 |
| Arizona | 25555.9 | 3970.3 | 0.0 | 29526.2 |
| Arkansas | 13775.5 | 0.0 | 0.0 | 13775.5 |
| California | 40564.1 | 534.6 | 6.5 | 41105.1 |
| Colorado | 17113.7 | 9846.0 | 0.0 | 26959.7 |
| Connecticut | 1299.7 | 0.0 | 0.0 | 1299.7 |
| Delaware | 416.7 | 112.8 | 0.0 | 529.5 |
| Florida | 15190.5 | 3.5 | 0.0 | 15194.0 |
| Georgia | 15257.7 | 0.0 | 0.0 | 15257.7 |
| Idaho | 18705.4 | 2937.9 | 0.0 | 21643.3 |
| Illinois | 14593.4 | 0.0 | 0.0 | 14593.4 |
| Indiana | 8479.8 | 892.3 | 0.0 | 9372.1 |
| Iowa | 8373.8 | 6201.6 | 0.0 | 14575.4 |
| Kansas | 12639.1 | 8670.8 | 0.0 | 21310.0 |
| Kentucky | 10466.1 | 0.0 | 0.0 | 10466.1 |
| Louisiana | 12367.7 | 0.0 | 0.0 | 12367.7 |
| Maine | 8615.7 | 0.0 | 0.0 | 8615.7 |
| Maryland | 2589.9 | 119.3 | 0.0 | 2709.2 |
| Massachusetts | 2145.6 | 0.0 | 0.0 | 2145.6 |
| Michigan | 12822.3 | 2336.3 | 0.0 | 15158.6 |
| Minnesota | 14650.6 | 7209.6 | 0.0 | 21860.2 |
| Mississippi | 12351.6 | 0.0 | 0.0 | 12351.6 |
| Missouri | 18051.7 | 0.0 | 0.0 | 18051.7 |
| Montana | 32331.1 | 5754.0 | 0.0 | 38085.1 |
| Nebraska | 14464.6 | 5570.5 | 0.0 | 20035.1 |
| Nevada | 27103.2 | 1532.2 | 0.0 | 28635.4 |
| New Hampshire | 2403.2 | 0.0 | 0.0 | 2403.2 |
| New Jersey | 1985.4 | 31.6 | 0.0 | 2016.9 |
| New Mexico | 19157.7 | 12335.0 | 0.0 | 31492.7 |
| New York | 12718.4 | 0.7 | 0.0 | 12719.0 |
| North Carolina | 13635.6 | 5.8 | 0.0 | 13641.4 |
| North Dakota | 12009.6 | 6302.4 | 0.0 | 18312.0 |
| Ohio | 10110.0 | 594.3 | 0.0 | 10704.3 |
| Oklahoma | 15349.3 | 2769.4 | 0.0 | 18118.7 |
| Oregon | 24243.4 | 898.6 | 0.0 | 25142.0 |
| Pennsylvania | 11734.9 | 0.0 | 0.0 | 11734.9 |
| Rhode Island | 314.0 | 0.0 | 0.0 | 314.0 |
| South Carolina | 8058.2 | 0.0 | 0.0 | 8058.2 |
| South Dakota | 14202.8 | 5770.2 | 0.0 | 19973.1 |
| Tennessee | 10915.3 | 0.0 | 0.0 | 10915.3 |
| Texas | 57787.8 | 11315.6 | 0.0 | 69103.4 |
| Utah | 18941.9 | 3047.1 | 0.0 | 21989.0 |
| Vermont | 2490.0 | 0.0 | 0.0 | 2490.0 |
| Virginia | 10463.5 | 95.1 | 0.0 | 10558.6 |
| Washington | 15305.6 | 2342.3 | 0.0 | 17647.9 |
| West Virginia | 6276.0 | 0.0 | 0.0 | 6276.0 |
| Wisconsin | 14363.3 | 180.4 | 0.0 | 14543.7 |
| Wyoming | 17770.7 | 7562.1 | 0.0 | 25332.8 |
| Total | 673553.5 | 108942.2 | 6.5 | 782502.1 |

**Appendix 4    Severity of Wind Erosion for Cultivated and Noncultivated Cropland by States in the Conterminous U.S. (Computed by Wind Erosion Equation)**

| State | <5.0 | ≥5.0 and <20.0 | ≥20.0 | Total |
|---|---|---|---|---|
| | | 1000 ha | | |
| Alabama | 1273.5 | 0.0 | 0.0 | 1273.5 |
| Arizona | 267.3 | 133.0 | 84.3 | 484.7 |
| Arkansas | 3128.2 | 0.0 | 0.0 | 3128.2 |
| California | 3983.1 | 56.6 | 28.2 | 4067.9 |
| Colorado | 1618.1 | 1458.2 | 541.7 | 3618.0 |
| Connecticut | 92.5 | 0.0 | 0.0 | 92.5 |
| Delaware | 198.5 | 3.5 | 0.0 | 202.0 |
| Florida | 1212.1 | 0.9 | 0.0 | 1213.0 |
| Georgia | 2093.4 | 0.0 | 0.0 | 2093.4 |
| Idaho | 1591.1 | 616.7 | 58.5 | 2266.3 |
| Illinois | 9752.9 | 0.0 | 0.0 | 9752.9 |
| Indiana | 5370.6 | 97.8 | 0.0 | 5468.4 |
| Iowa | 9451.6 | 658.1 | 2.6 | 10112.3 |
| Kansas | 9556.3 | 1121.5 | 72.7 | 10750.4 |
| Kentucky | 2060.6 | 0.0 | 0.0 | 2060.6 |
| Louisiana | 2416.6 | 0.0 | 0.0 | 2416.6 |
| Maine | 181.1 | 0.0 | 0.0 | 181.1 |
| Maryland | 677.1 | 0.0 | 0.0 | 677.1 |
| Massachusetts | 110.2 | 0.0 | 0.0 | 110.2 |
| Michigan | 3227.3 | 396.5 | 12.5 | 3636.2 |
| Minnesota | 5206.6 | 3181.2 | 254.6 | 8642.4 |
| Mississippi | 2317.3 | 0.0 | 0.0 | 2317.3 |
| Missouri | 5401.0 | 0.0 | 0.0 | 5401.0 |
| Montana | 3600.0 | 2183.8 | 300.6 | 6084.4 |
| Nebraska | 7162.9 | 583.5 | 39.5 | 7785.9 |
| Nevada | 249.1 | 29.6 | 29.7 | 308.5 |
| New Hampshire | 57.3 | 0.0 | 0.0 | 57.3 |
| New Jersey | 262.9 | 0.0 | 0.0 | 262.9 |
| New Mexico | 330.8 | 262.7 | 172.0 | 765.5 |
| New York | 2272.8 | 0.0 | 0.0 | 2272.8 |
| North Carolina | 2411.8 | 0.0 | 0.0 | 2411.8 |
| North Dakota | 8906.2 | 1023.3 | 83.8 | 10013.3 |
| Ohio | 4821.8 | 4.8 | 0.8 | 4827.4 |
| Oklahoma | 3667.1 | 377.6 | 34.8 | 4079.5 |
| Oregon | 1468.3 | 44.3 | 15.3 | 1527.9 |
| Pennsylvania | 2264.6 | 0.0 | 0.0 | 2264.6 |
| Rhode Island | 10.1 | 0.0 | 0.0 | 10.1 |
| South Carolina | 1207.0 | 0.0 | 0.0 | 1207.0 |
| South Dakota | 5710.8 | 920.3 | 20.5 | 6651.6 |
| Tennessee | 1965.5 | 0.0 | 0.0 | 1965.5 |
| Texas | 7001.3 | 2789.8 | 1646.0 | 11437.1 |
| Utah | 579.8 | 110.8 | 44.0 | 734.5 |
| Vermont | 256.8 | 0.0 | 0.0 | 256.8 |
| Virginia | 1172.3 | 1.0 | 0.7 | 1174.0 |
| Washington | 2121.0 | 486.6 | 122.0 | 2729.6 |
| West Virginia | 370.2 | 0.0 | 0.0 | 370.2 |
| Wisconsin | 4324.4 | 50.5 | 1.1 | 4376.0 |
| Wyoming | 566.6 | 172.0 | 180.6 | 919.3 |
| Total | 133948.3 | 16764.6 | 3746.5 | 154459.5 |

# Index